FINANCING RENEWABLE ENERGY PROJECTS

FINANCING RENEWABLE ENERGY PROJECTS

A guide for development workers

Jenniy Gregory, Semida Silveira,
Anthony Derrick, Paul Cowley,
Catherine Allinson, Oliver Paish

Practical
ACTION
PUBLISHING

Intermediate Technology Publications
in association with
The Stockholm Environment Institute 1997

Practical Action Publishing Ltd
25 Albert Street, Rugby, CV21 2SD, Warwickshire, UK
www.practicalactionpublishing.com

First published 1997\Digitised 2008

ISBN 10: 1 85339 387 8
ISBN 13 Paperback: 9781853393877
ISBN Library Ebook: 9781780444987
Book DOI: https://doi.org/10.3362/9781780444987

Since 1974, Practical Action Publishing has published and disseminated books and information in support of international development work throughout the world. Practical Action Publishing is a trading name of Practical Action Publishing Ltd (Company Reg. No. 01159018), the wholly owned publishing company of Practical Action. Practical Action Publishing trades only in support of its parent charity objectives and any profits are covenanted back to Practical Action (Charity Reg. No. 247257, Group VAT Registration No. 880 9924 76).

Reasonable efforts have been made to publish reliable data and information, but the author and publisher cannot assume responsibility for the validity of all materials or for the consequences of their use.

Acknowledgements
The review comments by the following people are gratefully acknowledged, as are the comments and suggestions received from colleagues of the authors:
Dr Michael Grupp, Synopsis, France; Lalith Gunaratne, Sri Lanka; Professor David Hall, Kings College London, UK; Mark Hankins, Energy Alternatives Africa, Kenya; Professor Bob Hill, Newcastle Photovoltaics Applications Centre, UK;Julie Smith, Enersol, USA; Nick Wardrop, Forum Secretariat, Fiji; Priyantha Wijesooriya, Solanka, Sri Lanka and Neville Williams, SELF, USA.
Thanks also to Jason Martin (UK) who assisted with the evolution of the first financing model used in this Guide.

Notice
Neither the publishers nor IT Power, nor the Stockholm Environment Institute make any warranty, expressed or implied, or assume any legal liability or responsibility for the accuracy, completeness or usefulness of any information, apparatus, product, method or process disclosed herein, or represent that it would not infringe privately owned rights. Reference herein to any specific commercial product, process or service by trade name, mark, manufacturer, supplier or otherwise, does not necessarily constitute its endorsement, favouring or recommendation by the authors or publishers of this Guide. The authors and publishers assume no responsibility for any personal injury or property damage or other loss suffered activities related to information presented in this book.

The manufacturer's authorised representative in the EU for product safety is
Lightning Source France, 1 Av. Johannes Gutenberg, 78310 Maurepas, France.
compliance@lightningsource.fr

Contents

Foreword

Renewable energy technologies provide one of the keys to a sustainable development future. The earlier these technologies are disseminated and put into broad commercial use, the sooner the world will approach a sustainable society for present and future generations.

Renewable energy technologies have been increasingly used in both industrialised and developing countries for different types of applications. In the past decades, growing environmental concerns and successful technological improvements have given new impulse to modern renewable energy technologies. Yet, there is still a huge gap between the technologies being readily available for use and their effective utilisation.

In general, it can be said that the process of dissemination of renewables has moved slower than is justifiable from a technological point of view in both industrialised and developing countries. The reasons vary from lack of knowledge among decision makers and users about existing alternatives, to policy and market failures which have favoured a supply-oriented approach to energy planning and continuous use of fossil fuels.

The provision of cheap, reliable and affordable energy remains a major challenge particularly to developing countries, as they strive to reach higher levels of economic development. However, increasing energy demand in these countries may lead to more pressure on natural resources and substantial environmental impacts both locally and globally.

Negative effects can only be reduced if sustainable practices for resource use are applied in combination with clean and renewable technologies to address energy needs according to local requirements, resource potential, and willingness and capacity to pay. This means going beyond conventional solutions which, besides not being sustainable, may not be readily justifiable from the economic point of view. It means also leap-frogging whenever possible so as to avoid unsustainable energy paths, and adopting renewable technologies that will guarantee reliable energy services at the least possible cost, the latter being based on a comprehensive cost calculation that includes equipment, life-cycle maintenance and environmental benefits.

In order for this to be possible, certain requirements need to be met. This includes policy measures that facilitate the penetration of renewable energy technologies in markets dominated by fossil-fuels, information dissemination and training. Financing mechanisms also need to be made available in order to secure international and local financial resources towards the acquisition of the desirable technology and its dissemination among a larger number of users.

Financing Renewable Energy Projects - a guide for development workers aims at addressing financing mechanisms for renewable energy technologies in developing countries, particularly the ones with very low income per capita in which the financing opportunities available to the poor are still very limited. While earlier guides produced by SEI and IT Power

FOREWORD

have been mainly focused on a number of energy technologies per se, this one addresses a decisive institutional issue in the concerted effort being made to improve the access of poor populations to electricity.

The Guide addresses financing mechanisms in a systematic way by showing their economics, applicability and advantages; warning about possible difficulties and shortcomings; and indicating the actors and institutions needed to operate them. The Guide is produced on the basis of proven experiences, and includes case studies to illustrate the variety of mechanisms able to be adapted to specific local conditions and needs. The overall objective of our initiative is to capitalise on successful experiences and contribute to destroying one of the main barriers that still hinders access to electricity for many millions in rural areas of developing countries. It is our hope that this material will cover a gap in the literature by treating the financial issue in an easy-to-read text format for practical use by NGOs, development workers, local communities, schools and donors who wish to start a financing scheme for small renewable energy systems in a developing region.

Semida Silveira
Stockholm Environment Institute
February 1997

Preface

This book is the result of co-operation between IT Power Ltd (UK) and the Stockholm Environment Institute (Sweden). The Stockholm Environment Institute (SEI) runs an information programme on renewable energy for development, which has resulted in a series of publications and seminars. This book is one of the results of this work. IT Power has substantial experience in the area of renewable energy projects in developing countries.

This book aims to provide information on different types of financing systems which can be applicable for small-scale renewable energy technologies, such as PV, small wind generators etc. As there are few current examples of such financing, it draws heavily from experiences with micro and small finance in other areas, such as micro-enterprise and agricultural lending. In giving many examples and possible financing structures, a model on how to make a fund revolve and highlighting sources of financing and technical issues to be aware of, it is hoped that this book will assist in promulgating financing options for renewable energy technologies in developing countries.

This is the fifth in the series of guides for development workers on energy technologies for sustainable development. Other guides are:

Solar Photovoltaic Products
First ed. January 1989 (A Derrick et al)
Updated January 1991
ISBN 1 85339 091 7

Micro-Hydro Power
January 1992 (P Fraenkel et al)
ISBN 1 85339 029 1

Windpumps
April 1993 (P Fraenkel et al)
ISBN 1 85339 126 3

Rural Lighting
November 1993 (J P Louineau et al)
ISBN 1 85339 200 6

Contact persons:

Jenniy Gregory
IT Power Ltd
The Warren
Bramshill Road
Eversley
Hampshire RG27 0PR, UK

Fax: +44 118 973 0820
email: jag@itpower.co.uk

Semida Silveira
Stockholm Environment Institute
Box 2142
103 14 Stockholm
Sweden

Fax: +46 8 723 0348
email: semida.silveira@sei.se

Introduction 1

Lack of energy infrastructure remains a serious constraint to the development of many developing countries. Together with transport and communication systems, an effective energy system contributes to increasing efficiency in traditional production activities, establishing new industries and diversifying the economy, thus contributing to economic development.

Access to affordable energy can be considered as an essential component of any sustainable socio-economic development strategy. It is fundamental for the improvement of the living standards of populations that lack access to reliable and efficient energy, particularly electricity. In fact, over 2 billion people do not have access to electricity even to fulfil basic needs.

Meeting the energy needs of large populations involves mobilisation of considerable sums of capital for investment in energy infrastructure. But while infrastructure needs have increased rapidly with population growth, urbanisation and economic development, sources of capital for large scale investments have become more competitive and scarcer.

However, because of its central importance in the development process, the size of initial obstacles to be crossed, and the complexity of energy systems, the energy sector continues to receive substantial attention from national governments, donors and development agencies. Meanwhile, investment scarcity and the absence of large established energy markets

in developing countries have led to a gradual shift from the conventional view of simply building large centralised energy systems, to facilitating access to electricity through small and decentralised systems. It is believed that this new approach is more flexible, and will speed up the electrification of many developing regions.

Although fossil fuels are still by far the dominant source of energy in the world, renewable energy systems have been increasingly applied in both small and large scales in industrialised and developing countries. In many developing countries, biomass still dominates as the main energy source though, traditionally, it has not been used for electricity generation.

The most common modern form of renewable energy technology for electricity production has been large hydro power plants. However, during the last decades, solar power, wind and biomass have become attractive alternatives for both small and large scale electricity production while large hydroelectric power schemes have been increasingly criticised for the large social and environmental impacts that they normally cause.

Small renewable energy systems are very attractive for many developing countries. They provide the least cost electricity service in isolated rural areas particularly in areas that are not connected to the electricity grid. These systems include solar photovoltaic (PV), micro-hydro, wind power and biogas.

1

Solar water heaters, solar stills, solar cookers, biomass for efficient cookstoves and wind pumps are other often cost-effective renewable sources with relevant applications in both rural and urban areas.

Benefits of renewable energy technologies

The environmental benefits of renewable energy systems are generally not yet costed into the energy equation. Because some renewable energy technologies have high initial capital costs, electricity generated from these sources is often wrongly seen as being more expensive than electricity generated by fossil fuels and nuclear power.

This is mainly due to narrow cost analysis. Issues such as water and ground pollution, greenhouse gas emissions, decommissioning costs, environmental degradation, hazardous working conditions, cost of importing fuel, etc. are not part of the fossil fuel and nuclear energy equation. The prices of diesel generating sets do not include the future fuel consumption and their environmental costs.

The use of most renewable energy systems is not detrimental to the environment and often has no net emissions of carbon dioxide, except in the unsustainable use of biomass resources. From the overall perspective of development, there is no doubt that renewable sources of energy will form the basis for a sustainable future, guaranteeing not only a balance between development and natural resources use but also self-reliance and strengthening of local economies.

The global environmental benefits of renewable energy technologies have been pushed forward particularly by the United Nations Framework Convention on Climate Change. Major shifts in the global energy matrix are required to comply with the objectives of the Convention. A gradual move away from fossil fuels and towards more renewable energy sources is one of the major keys to meet this challenge.

Other global agreements such as the Agenda 21 and the Biodiversity Convention also impose new challenges on society and have contributed to change the context in which

energy systems operate. Therefore, financial and institutional mechanisms have been created which affect the views on how energy supply and demand are to be resolved. This has had a positive effect in terms of improving acceptance and the conditions for dissemination of renewable energy technologies.

Thus, the challenge is not simply to meet energy demands and promote development. Rather this has to be accomplished on a sustainable basis, which means that the energy systems established will have to be efficient in terms of production and use, the technologies will have to be reliable, they will have to be environmentally sound, institutionally manageable and economically sound.

Besides the overall global benefits, renewable energy systems can provide direct benefits at national and local levels which justify their wide use in developing countries. They can contribute to substantial savings in import bills for fossil fuels, which are a main drain in the economy of these countries. At the local level, the availability of electricity will contribute to improved productivity in existing cottage industries and create opportunities for the establishment of new ones. Indirect positive effects include the provision of jobs through local manufacturing, installation and maintenance of the equipment.

Other examples of benefits of renewable energy technologies to both the community and individuals include:

- education (through the provision of lighting and TV)

- potable water (water pumping, solar stills and desalination)

- increased opportunities for growing crops (water pumping)

- reduced labour foraging for fuel (improved cookstoves and solar cookers)

- increased health services (PV lighting and refrigeration, solar water heaters and solar stills)

INTRODUCTION

Organisation of the Guide

There are a number of issues that affect the wide-scale dissemination of renewable energy technologies in developing countries. These can be technical, institutional or financial. One of the main obstacles is access to appropriate financing which this Guide proposes to help remove.

The Guide looks at the issues which affect the success of a renewable energy programme particularly with regard to financing. This includes sources of funds, alternative mechanisms for financing renewable energy equipment to low-income users, and institutional, implementation and management issues. Some technical aspects are also mentioned, since it is believed that those interested in setting up financing schemes for renewable energy technologies must also be aware of the technical parameters which need to be factored into any scheme.

In looking at renewable energy technologies, PV has been chosen as the primary example to be covered here, since it is one of the most cost-effective ways of bringing electricity to remote areas. With batteries for storage and no fuel requirements other than sunlight, PV systems can power lights and TV in the home, provide electricity for small enterprises, employment for local technicians, and general improvement in welfare. They are easier than other systems to finance, as they have a long life-time and thus residual value, and are modular and thus can be added to or dismantled and sold. Moreover, there are thousands of solar systems already installed around the world which indicates also that the technology is well tested and accepted.

Many of the examples given in this book and the financing issues discussed are applicable to other renewable energy technologies. Grid-connected systems such as wind farms, and minigrids such as those powered by micro-hydro systems are not considered here due to their different financial and technical requirements.

There are many technical books which can provide more information on technical matters and books which relate experiences from micro-credit schemes for other applications such as agriculture and micro-enterprises. The aim of this publication is to act as an initial reference and a guide for development workers who wish to finance small renewable systems.

Chapter 2 provides a brief overview of the types of technologies that can be of interest for developing countries. Readers who are not very familiar with these technologies may need to consult other sources for more information. Some of these technologies have been treated in detail in other SEI and IT Power Guides.

Chapter 3 demonstrates economic assessments and comparisons that should be made when different technologies are being contemplated for a scheme. The chapter is complemented by the life-cycle costing analysis provided in **Appendix 1**. This type of information helps build a strong argument, which can be particularly useful in discussions with policy makers and economists.

Chapter 4 gives an overview of issues that have to be kept in mind when attempts are made to disseminate renewable energy technologies. These issues are further discussed in other chapters starting with the roles of international, national and local institutions which are addressed in **Chapter 5**.

The type of financing scheme chosen and the source of funding for it affect the type of services that can be offered by the scheme and its long-term sustainability. **Chapters 6, 8 and 9** discuss the different financing mechanisms and the parameters for setting up a revolving fund. This information is the key in the whole of the Guide and is complemented by **Appendix 2**, which contains assumptions on a working model of a revolving fund.

Chapter 7 looks at sources of funds, and **Chapter 10** provides a check list for implementing a fund.

Five case studies are also provided at the end illustrating the successful experiences of Dominican Republic, Kenya, Sri Lanka, Democratic Republic of Congo (formerly Zaire) and Zimbabwe with the dissemination of PV systems. Kenya has around 40,000 Solar Home Systems installed, the majority of

3

which have been bought for cash. In Sri Lanka and the Dominican Republic, NGOs have played a major role in the dissemination process not least by establishing financing activities. The national health programme from 1988-91 is the example in the Democratic Republic of Congo. In Zimbabwe, it is the Global Environment Facility PV programme that is reviewed. The case studies reinforce the notion that committed institutions and individuals, secured funds and good technology are essential requirements which have to be adapted to the particular characteristics of the targeted group of users

if renewable technology schemes are to be successful.

Finally, it may be worth saying that the Guide is of most relevance for those who work in low-income countries and areas. Some developing countries have well established financial institutions and markets in which case some of the schemes discussed may be of less interest. Yet, middle-income countries often have large income disparities among their populations and peripheral and remote communities in these countries may find the information provided here of great use.

Renewable Energy Technologies   2

The majority of the people in developing countries live in rural areas, without the energy services which are taken for granted in towns and cities. Most areas are unlikely to be connected to the electricity grid in the foreseeable future.

However, renewable energy technologies (RETs) can provide good quality alternatives, and indeed, can be more reliable in providing electricity than grid extension (Hulscher & Fraenkel, 1994). There are significant numbers of economically-attractive applications for renewable energy technologies today (WEC, 1993). Different technologies will suit different regions. This will be influenced by culture and climate (there is no point having a wind generator if there is little or no wind). These systems will typically be what are known as 'stand-alone' systems, i.e. unconnected to any centralised electricity grid.

There are a number of factors which all the RETs have in common. These include:

- there are both good quality and inferior products available on the market. Care must be taken when selecting equipment that there are warranties and the products are good quality.

- all equipment requires maintenance, even though this may be minimal.

- there must be an adequate supply of spare parts, plus trained technicians available to undertake maintenance and repair.

The technology which has most universal application for providing least cost electricity to rural areas of developing countries is photovoltaics (PV). It can be used for a variety of power applications and has many additional benefits when compared to other technologies. There is a mature market to support PV world-wide, and systems can be manufactured locally, either in part or totally (EPIA, 1996).

Many of the financing schemes and systems described in this book refer particularly to small household PV systems, known as Solar Home Systems (SHSs). These stand-alone PV-powered home lighting kits usually also are large enough to power a TV, and are typically around 30 - 50 Wp in size. Many thousands of these have already been installed world-wide. There are three major reasons for this choice of renewable energy application:

- SHSs bring electricity to homes, work, schools, health centres, agriculture etc. where the alternatives are either more expensive or non-existent.

- SHSs and PV for small enterprise development can be financed using traditional financing structures.

- There are already a number of schemes in operation in developing countries to support the installation of such systems.

5

2.1 Photovoltaics

PV systems convert sunlight (both direct and diffuse) into electricity. The building block of such systems is the PV module. There are two types of modules on the market: those made from crystalline silicon cells, and those made of thin-film materials. Crystalline silicon modules have a long lifetime and do not degrade in efficiency. They are usually more expensive than the current thin-film technology, amorphous silicon. Amorphous silicon modules have a lower efficiency, do degrade over their lifetime, but are generally less

expensive than crystalline silicon modules. There is intensive research and development underway to bring new thin-film PV modules onto the market.

There are a number of advantages which PV systems have:
- cost effective in many applications
- reliable
- low maintenance
- environmentally benign
- abundant fuel
- local-generated power
- flexible size
- transportable
- long lifetime
- modular
- local manufacture of components

However, there are limitations:
- Sun dependent
- High initial cost
- System maintenance
- Power type (DC)

PV modules have been used for years in a variety of applications: telecommunications, railway signalling, street lights, water pumping, solar home systems etc. They have a good track record, particularly for professional systems, such as powering remote telecommunications transmitters where they are considered the only reliable and low maintenance power source for remote areas (Derrick et al, 1991).

Solar Home Systems

A SHS can provide low-power electricity, and systems are usually sized to provide the minimal power needed by two or three fluorescent lamps and a radio/stereo or television set. SHSs are designed to serve small energy users. Larger Remote Area Power Systems (RAPS) can cater for larger electricity demands.

SHSs are cost competitive with electrification when serving disperse, low-energy demand households. Systems are modular, enabling capacity to be upgraded as needed. Little maintenance is required, but batteries need replacing every three to five years.

The electricity produced by PV modules is Direct Current (DC), so an inverter is neces-

---

Table 2.1 Applications for PV by sector

- Agriculture
- water pumping
- electric fencing for livestock and range management

- Small enterprises
- lighting systems, to extend business hours and increase productivity
- power for small equipment, such as sewing machines, freezers, grain grinders, battery charging
- lighting and radio in restaurants, stores and other facilities

- Healthcare
- lighting for wards, operating theatre and staff quarters
- medical equipment
- refrigeration for vaccines
- communications (telephone, radio communications systems)
- water pumping
- security lighting

- Community
- water pumping, desalination and purification systems
- lighting for schools and other community buildings

- Domestic
- lighting, enabling studying, reading, income-producing activities and general increase in living standards
- TV, radio and other small appliances
- water pumping

sary if Alternating Current (AC) electricity is required (IT Power, 1996a).

PV pumping

PV pumping systems can be used for community water, stock watering and irrigation. The size of the system depends on the difference between the level of the water at the source (well, river etc.), and the level to which it must be lifted (the head). There are many examples of PV community water supply systems, particularly in the Sahelian region of Africa, where PV boreholes provide clean and reliable services. PV pumps have also been developed for small-scale low-head irrigation, but these have not yet been used in large numbers (Barlow et al, 1993).

PV for healthcare

There are many thousands of PV-powered refrigerators in clinics and hospitals in the developing world. These keep vaccines cool and complement the immunisation programmes. PV can also be used to power lights for the operating theatre and hospital wards and to provide power for the staff quarters.

As is shown in the case study on Zaire, PV installations can also be used to raise revenue for the hospital or clinic. During the day, for instance, if the clinic's batteries are fully charged, the excess electricity can be sold to

PV pumping, Tionbiogou, Mali   Photo: IT Power

recharge other batteries. Excess electricity can be used to power community videos. Other solar technologies are also beneficial for use in clinics and hospitals, such as solar water heaters, solar sterilisers and passive solar design for natural cooling, heating and ventilation (Zaffran, 1993).

2.2 Solar thermal electric systems

Electricity can also be generated using solar-powered heat engines. Solar collectors heat the working fluid of a heat engine connected to a generator. Heat engines operate more efficiently at higher source temperatures. To obtain economically viable systems, most solar thermal systems use concentrating solar collectors.

Installation of a PV system on a health clinic in Eritrea   Photo: Dulas

Solar water heater, China   Photo: IT Power

Large-scale grid-connected solar thermal electric systems have been commercialised, but small-scale, stand-alone systems are not yet commercially available. The large-scale systems are not for individual use, but could be applied in a community power project or mini-grid. Although many working prototypes of small-scale solar thermal electric systems do exist, none have yet been proven to compete with the reliability and modularity of solar photovoltaics (EUREC Agency, 1996). Such systems are not considered appropriate for small-scale financing at a local level.

2.3 Solar water heaters

Solar water heating (SWH) systems heat water using solar collectors. There are a number of well-developed technologies, primarily close-coupled thermosyphon flat plate collectors with header tanks, and evacuated tube technology. There are also simpler systems, which are less efficient but which are cheaper and meet the needs of many people in less developed areas. There are millions of SWHs installed world-wide (ESIF, 1996).

SWHs have a number of technical advantages: they can be designed to avoid the need for moving parts, are very robust, require little maintenance, and stand up to much wear and tear. They are relatively easy (though bulky) to transport, and require a minimum of expertise to install. In most cases, local plumbers are able to install the systems. System lifetimes are generally above ten years, depending on their sophistication (EUREC Agency, 1996).

2.4 Solar dryers

Open-air drying has been used for centuries to dry fruit and vegetables, but developing countries regularly suffer heavy post-harvest losses of food using this traditional drying method. The use of solar air collectors to pre-heat air circulating through a drying chamber reduces these losses and is more predictable than the open air alternative.

Solar dryer technologies today can either use natural convection to pass warm air through the collector or use a fan to create air flow. In most cases, air is heated in a collector, then

Solar drying of coffee, cacao and copra in Indonesia   Photo: Innotech

Solar box cooker, Pniel, South Africa   Photo: Synopsis

2.6 Solar cookers

There are a number of solar cooker technologies on the market. These are variations of boxes, concentrators and flat plate cookers with and without storage. These use the direct heat from the sun to cook the food placed within them. Solar cookers vary in both price and construction detail. Some are made from very simple materials (such as cardboard and aluminium foil), others from iron and glass. There are also more complicated communal systems available (Hulscher & Fraenkel, 1994).

passed through the grain, fruit or vegetables to be dried. Forced convection systems using a fan can reduce drying time by a factor of three; however, they are more expensive to install. Even the simplest designs are inexpensive in relation to the yearly crop losses and thus the availability of appropriate financing is beneficial. Solar dryers can be used by individuals, families, farmers or co-operatives (Hulscher & Fraenkel, 1994).

There have been a number of relatively large dissemination activities undertaken in countries such as India, China and Kenya. In South Africa, a new comparative field test programme is evaluating the different cooker models.

2.5 Solar stills

Simple solar distillation technologies that evaporate saline or polluted water and recondense it as pure distilled water, are well established. The most common type of system is the single-basin still. Multiple-effect stills have higher efficiencies, but also higher costs and complexity.

In wick stills, water flows slowly through a porous, radiation-absorbing pad (the wick). The pad can be tilted to give a better sun angle and, because there is less water in the basin, it can be heated more quickly. (ECSCR, 1994)

2.7 Improved cookstoves

During the last 35 years, large efforts have been put into improving traditional cookstoves and disseminating them in countries of

Biogas digester, Pura, India   Photo: D O Hall

9

Commercial biomass system, Kristva, Sweden   Photo: D O Hall

Africa and Asia. Many programmes have been successful such as in India, China and Kenya. Almost all countries in Eastern and Southern Africa have improved stove projects either at the national or grassroots level.

Initial efforts to find a universally accepted concept and alternative have failed because local differences in terms of fuel availability and traditions were not considered. Differences are also found between urban and rural stoves. While urban stoves are usually single-fuelled and households obtain fuel

in a market, rural stoves often use various fuels which are usually collected rather than bought.

In general, improved stoves save fuel because the heat loss is minimised through some type of insulation. But they may also be more expensive than the traditional kind. Nevertheless, the improved stoves are normally produced locally and have substantial positive impacts on women's health and labour intensity, which implies social benefits beyond the energy savings accrued (Karekezi & Mackenzie, 1993).

2.8 Biomass and biogas energy technologies

Biomass is currently the world's fourth largest energy source, contributing 14% of the world's primary energy demand and 35% in developing countries (EUREC Agency, 1996). The most important sources of biomass for energy are residues from crops, animals and forestry industries, wood resources from natural forests and managed plantations. Biomass energy technologies vary from the most traditional to modern sophisticated ones. For example, many developing countries utilise biomass in traditional industries such as brick and lime kilns.

Large-scale biomass plants for electricity or cogeneration are becoming increasingly more common. In the last 15 years, Sweden has invested in a large programme for increasing the use of biomass for energy. Crop and saw-mill residues and energy forests are the main source. The same potential exists in many developing countries and remains largely untapped. Valuable saw-mill residues,

Kijito windpump, Kenya   Photo: IT Power

10

bagasse and other crop residues are often burned when they could be used to produce heat and electricity, adding economic value to industries. Surplus electricity can be sold to the grid benefiting local communities.

Donors and development agencies are becoming increasingly aware of the large potential of biomass that exists in many developing countries for electricity production. As the monopolies of utilities collapse in many countries, the private sector is likely to become an important actor in utilising the existing potential.

Wind farm, Delabole, UK   Photo: IT Power

The Global Environment Facility (GEF) is investing in a large-scale pilot project in Brazil to demonstrate the technology to generate electricity through plantation-grown fuel. It is also assessing their commercial feasibility and environmental compatibility.

Biogas is produced by the anaerobic fermentation of biomass such as dung, municipal wastes and crop residues. Diesel engines can be converted to burn biogas and generate electricity or to operate on a dual-fuel (biogas-diesel) mode. The technical options available vary from simply sinking pipes to extract gas from existing landfills to large-scale digesters. Agricultural and forestry residues available in rural areas of developing countries provide the necessary input for this technology. In India, biogas programmes have been carried out using animal wastes.

Biogas plants can be reliable and simple to maintain though care has to be taken to avoid leakage of methane and explosion risk. The fuel is locally available and cheap indigenous technology can be

applied making rural electricity systems self-reliant. The plants can be family-sized though, in this case, their output is usually more suitable for cooking than for generating electricity. Community plants are more economical but the maintenance and distribution of costs and benefits are more difficult to organise (Wereko-Brobby and Hagen, 1996).

Biogas technologies are being disseminated in countries like India and China using small-scale locally produced digesters. The plants can be designed to fit the minimum input

Small wind generator   Photo: Marlec

11

available or maximum gas consumption required (Rajabapaiah et al, 1993). In Tanzania, GEF is establishing the first large-scale biogas plant in Africa, converting methane from municipal waste into energy. Large-scale systems are not considered appropriate for small-scale financing at a local level.

Ethanol has become a relevant alternative fuel in the transport sector not least because of the considerable environmental benefits that can be obtained. Many industrialised and developing countries have initiated ethanol programmes or projects during the last years, the most comprehensive being the Brazilian fuel-alcohol programme started in 1975. There is also a large-scale alcohol-gasoline programme in the USA, and other countries are conducting R&D and pilot programmes for producing bioethanol (Murphy et al., 1996).

2.9 Wind pumps

There are over 1 million wind pumps in operation around the world. Their function is to pump water, using wind energy. Wind pumps do not produce electricity. This mechanical technology is relatively simple to manufacture, operate and maintain. Some wind pumps are designed for high wind situations, others for low wind thresholds (Barlow, Bokalders et al, 1993). Care must be taken to choose the right wind technology for the local wind regime.

Wind pumps can be manufactured in small engineering workshops in most developing countries. In some developing countries (e.g. Zimbabwe and Kenya) wind pump manufacture is also an export industry.

2.10 Wind generators

Wind generators (or turbines) use the power of the wind to generate electricity. They generate power cost-effectively in areas with a good wind resource. Wind generators vary in size, and fit into three categories: medium to large-sized grid-connected machines (150-1500 kW in size); intermediate sized turbines (10 - 150 kW in size range); and small stand-

alone turbines (< 10 kW in size) (EUREC Agency, 1996). The nomadic people of Mongolia have used licensed locally manufactured small wind turbines to power lights, radios and television for many years. Whilst initial capital costs are high, there is no fuel and little maintenance is required.

Medium and large-sized wind turbines can be connected to the electricity grid or provide power for stand-alone applications, such as fish processing plants. They are a commercial technology, and many are being installed in India and Egypt to strengthen the grid. Such systems are not considered appropriate for small-scale financing at a local level.

2.11 Micro hydro power

The basic physical principle of hydro power is that if water can be piped from a certain level to a lower level, then the resulting water

Micro hydro system in Nepal, used for crop processing and electricity generation   Photo: O. Paish

pressure can be used to do work. If the water pressure is allowed to move a mechanical component, then that movement involves the conversion of water energy into mechanical energy. Hydro-turbines convert water pressure into mechanical shaft power, which can be used to drive an electricity generator, a grain mill, or some other useful device (Fraenkel et al, 1991).

The main advantages of micro and mini hydro power are:

• power is usually continuously available on demand

• given a reasonable head, it is a more concentrated energy resource than wind, solar or biomass

• the energy available is largely predictable over the year

• no fuel and only limited maintenance are required, so running costs are low (e.g. compared with diesel power) and in many cases imports are displaced to the benefit of the local economy

• it is a long-lasting and robust technology

• systems can readily be engineered to last for 50 years or more without the need for major new investment

Against the advantages, the main shortcomings of micro hydro power are that it is a site-specific technology; and sites that are well suited to the harnessing of water power are not always close to a location where the power can be economically exploited. There is always a maximum useful power output available from a given hydro power site, which limits the level of expansion of activities which make use of the power. And finally, river flows often vary considerably with

Table 2.2 Applications for renewable energy technologies, by sector

TECHNOLOGY	SECTORS					
	Domestic	Small business	Agriculture	Health	Community	Other
Solar energy						
PV	•	•	•	•	•	•
solar thermal electric		•			•	•
solar water heating	•	•		•	•	•
solar cookers	•	•			•	
Biomass						
improved cookstoves	•					
briquetting	•	•				•
district heating					•	•
electricity generation						•
Biogas						
electricity generation	•	•		•	•	•
transport fuels		•				•
Micro/small hydro		•			•	•
Wind						
turbines	•	•		•	•	•
pumps	•		•	•	•	

the seasons, especially where there are monsoon-type climates, and this can limit the firm power output to quite a small fraction of the possible peak output.

2.12 Stand-alone systems versus Mini-grids

The preferred option for rural electrification in developing countries, by most households, is connection to the main grid. However, the cost of grid extension is prohibitively expensive in many cases, and stand-alone renewable energy technologies, such as PV, are more cost effective.

However, there is another option, and that is to install a 'mini-grid'. This is a small electricity grid system which connects a small number of users (e.g. 5 - 200), and which is powered by a local electricity generator. To date, most of these systems are powered by diesel generators or micro-hydro. However, PV or wind-powered mini-grids can also take advantage of the economies of scale available through serving many customers.

Despite their clean energy conversion and little maintenance costs, pilot PV and wind mini-grids have generally not performed well to date in developing countries. The principal problem appears to be that users connected to this grid expect to be able to use everyday 120/240V AC appliances. If there are no electricity meters per user, and electricity use is not rationed (i.e. used in accordance with what the system can produce), then some users will consume large amounts of electricity leaving small amounts available for oth-

ers. A more successful option for a mini-grid may require a sophisticated variable current limiting device, or that it be run as part of a hybrid system, for example PV/diesel (IT Power, 1996b).

Mini-grid systems are often more expensive to establish than individual distributed systems and are more appropriate for utilities and Energy Service Companies (ESCOs). They thus require different financing than SHSs or other smaller-scale RETs, and, for that reason, are not discussed in the remainder of this book.

2.13 Conclusions

There are many different types of renewable energy technologies available on the market which can bring electricity to villages and communities which do not have access to grid electricity. Each of these systems has advantages and disadvantages, and each must be assessed in relation to the resource available to power it, so that the most appropriate systems are installed for the climatic conditions. Likewise, as the following chapters will show, systems must be assessed as to their costs, economics and suitability for the local people. Table 2.1 outlines the different applications for PV systems by sector. Table 2.2 outlines many of the different applications for renewable energy technologies, by sector.

The following chapter looks at the costs and economics associated with stand-alone applications, particularly PV installations.

Costs and Economics 3

3.1 Overview of equipment costs

This section provides an overview of the typical costs of different types of stand-alone energy equipment. It must be noted that equipment prices, in what is still a small and immature market for renewable energy technologies, can be highly variable, depending on the country and location, the source and quality of equipment, and the taxes or subsidies applied.

Table 3.1 therefore provides only indicative costs for a limited range of hardware, based on the authors' experience of the renewable energy market. Where possible these are listed as a cost per unit of installed capacity (e.g. $/Wp for solar systems).

Figure 3.1 shows a more detailed breakdown of the cost of a domestic PV lighting kit (Hankins, 1995). The chart illustrates the relative costs of different system components, and in particular the considerable proportion of system cost that can be attributed to import duties and purchase tax.

Cost reduction

There are prospects for future cost reductions as the different markets develop. For example, the price of PV modules has reduced

Table 3.1 Costs of renewable energy systems

Equipment	Size range	Cost Guidelines
PV modules	50 Wp or greater	$5-7/Wp
PV lighting systems	50 Wp or greater <50 Wp	$10-15/Wp $15-30/Wp
PV water-pumping systems	100-1000 Wp	$10-15/Wp
Micro-hydro: electro-mechanical equipment	30-300 kW at heads >50 m 30-300 kW at heads 10-50 m	$400-600/kW $600-1200/kW
Micro-hydro: complete installations	30-300 kW at heads >50 m 30-300 kW at heads 10-50 m	$1500-2000/kW $2000-4000/kW
Wind-generator and battery unit	from 100 to 10,000 W at 10 m/s wind-speed	$3-5/W
Wind pumps (excluding borehole and water tank)	from 1.8 m to 7.4 m rotor dia- meter (2.5-43.0 m² rotor area)	$150-350/m² of rotor area

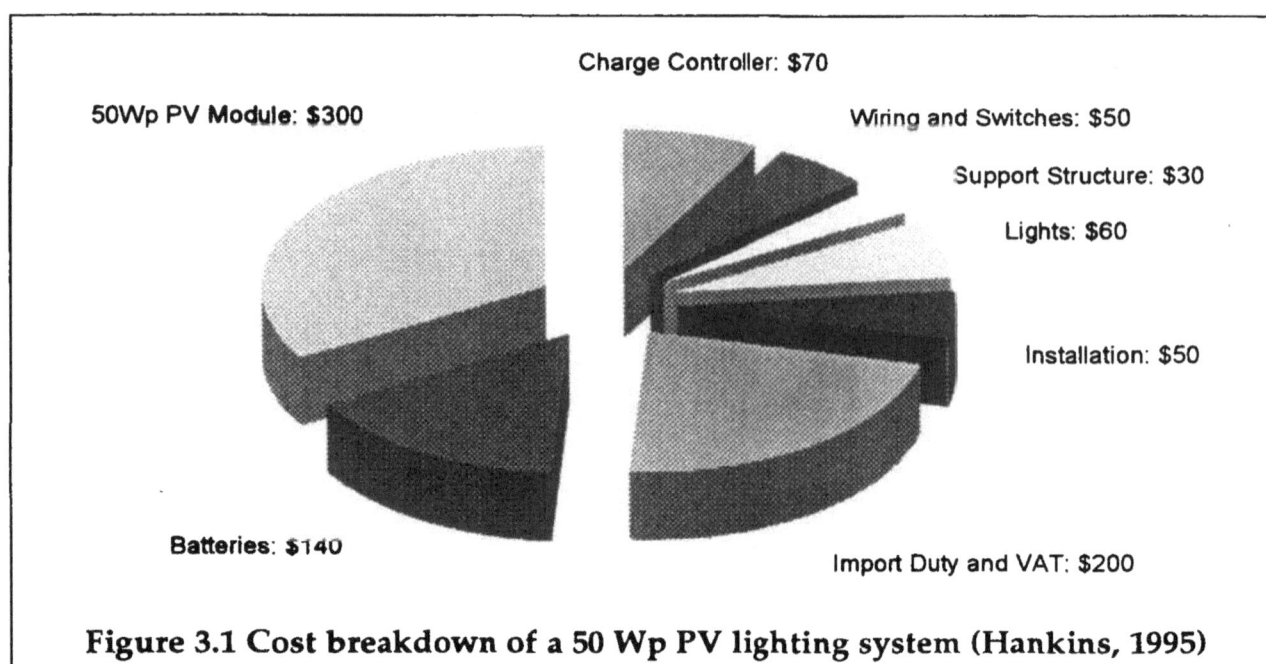

Figure 3.1 Cost breakdown of a 50 Wp PV lighting system (Hankins, 1995)

from \$20/Wp to around \$5/Wp in the period 1980-95 and is anticipated to fall by at least a further 50% by 2005.

The cost of micro-hydro schemes can be highly variable anyway, depending upon where the equipment is manufactured and how much civil engineering is required to implement the scheme. The cheapest schemes are those where the equipment is manufactured locally and the majority of the civil works are completed by the owner with the help of local contractors. Therefore the biggest scope for cost reduction is in increasing the manufacturing capability in the country concerned so as to avoid imported hardware. The same applies to windpower equipment, where joint ventures between Western and developing country manufacturers are bringing manufacturing costs down, assisted by a general expansion of the market.

3.2 Economic assessments

Introduction

The initial cost is only one element in the overall economics of a system. Some type of economic assessment is required to determine which system from a number of choices will give the best value for money in the longer run, either for the customer or for the economy as a whole.

The economics of renewable energy technologies are rather different from conventional small power systems, in that:

- the capital cost of the equipment is high

- the running costs are low and there are no fuel costs

- the output of the system depends on its location

In particular it is the high capital cost of renewables which makes them appear unattractive when compared with a diesel generating set, and makes an appreciation of the longer-term economic picture so important. This is normally achieved through the method of life-cycle costing.

Life-cycle costing

Appendix 1 summarises a widely-used methodology for analysing the value-for-money (or cost-effectiveness) of renewable energy systems relative to each other, or relative to conventional power systems. It examines all the costs incurred over the lifetime of different systems, and compares them on an

16

COSTS AND ECONOMICS

Table 3.2 Useful technical data

Power and energy	• 1 watt (W) of power consumes 1 watt-hour (Wh) of energy in 1 hour • 1 kilowatt (kW) equals 1000W. 1 kWh=1000 Wh
Photovoltaics	• 1 Wp of PV module capacity produces 1 W of electrical power in 'peak' sunshine of strength 1000 W/m². • 1 Wp of PV produces 1 Wh per day of electrical energy in an average solar insolation of 1 kWh/m²/day.
Wind	• The gross power passing through an area Am² in a wind of speed Vm/s is $0.6AV^3$ Watts. • The mean efficiency of a small wind-generator is around 25%. • Hence the electrical power captured by a rotor of area A in a mean wind speed of V is about $0.15AV^3$.
Hydro	• The gross power available in a flow of Q litres/sec falling through a height Hm is $9.81QH$ Watts. • A micro-hydro plant converts typically 50-70% of this power into electricity.
Diesel	• Energy content of diesel fuel: 10 kWh/litre. • Conversion efficiency of a small diesel engine: 15-25%.
Petrol / Gasoline	• Energy content of petrol fuel: 9 kWh/litre. • Conversion efficiency of a small petrol engine: 6-10%.
Batteries	• A 100 ampere-hour (Ah) 12 volt (V) battery stores 100 x 12 = 1200 Wh of electrical energy. • Ordinary batteries should not normally be discharged by more than 30% of their storage capacity. Deep discharge batteries can be discharged by 50% or more. • Batteries can be charged at a rate of C/10 kilowatts (C=kWh capacity) without damage. • Batteries can be assumed to have a storage efficiency of 70-80%. • Good solar batteries cost in the range $150-200 per kWh of stored energy.
Grid Extension	• The cost of extending an 11 kV grid is usually between $10,000/km and $15,000/km.
Capacity Factor	• Capacity Factor is the ratio of the maximum achievable energy output per year to the maximum output if the system were able to run at full rated capacity all year round. • A 1 kW system with 10% capacity factor produces 876 kWh per year.
Typical Capacity Factors:	• PV 10-20% • Wind 20-40% • Micro-Hydro 60-90%
Load Factor	• Load Factor is the ratio of the energy consumed per year to the maximum output if the system were able to run at full rated capacity all year round. An oversized system will have a low load factor. • The Load Factor must be less than or equal to the Capacity Factor.

equal basis by converting all future costs into today's money. This method is known as a discounted life-cycle cost analysis.

For instance, a PV array costs more to buy than a diesel generator, but the modules should last over 20 years, while the diesel generator might last 10 years. In this case the analysis period would be 20 years. In addition to the capital cost, the cost of a replacement diesel after 10 years, plus 20 years' worth of fuel are also included in the cost of the diesel option. The costs of maintenance and repair for the two systems over the whole 20 year cycle must also be added. Depending on the exact figures, either the PV or the diesel system will work out cheaper overall.

Life-cycle costing enables one to appreciate how the various costs involved over a period of time can be simplified into a fixed annual or monthly cost. This can then be used as a basis for determining the regular payments required within a financing scheme.

Basic technical data needed to make comparisons between the different technologies are summarised in Table 3.2.

Economic comparisons

Different energy technologies prove to be more cost-effective in different situations. In general, it is the average amount of energy required per day which dictates the choice of system. In the examples below, life-cycle costing has been used to compare the net cost (or levelised cost) of energy supplied by different technologies over their lifetime, as a function of the daily energy required.

Levelised energy cost is equivalent to the income (in today's money) that would have to be received for each unit of energy consumed in order for the project to break even over its lifetime.

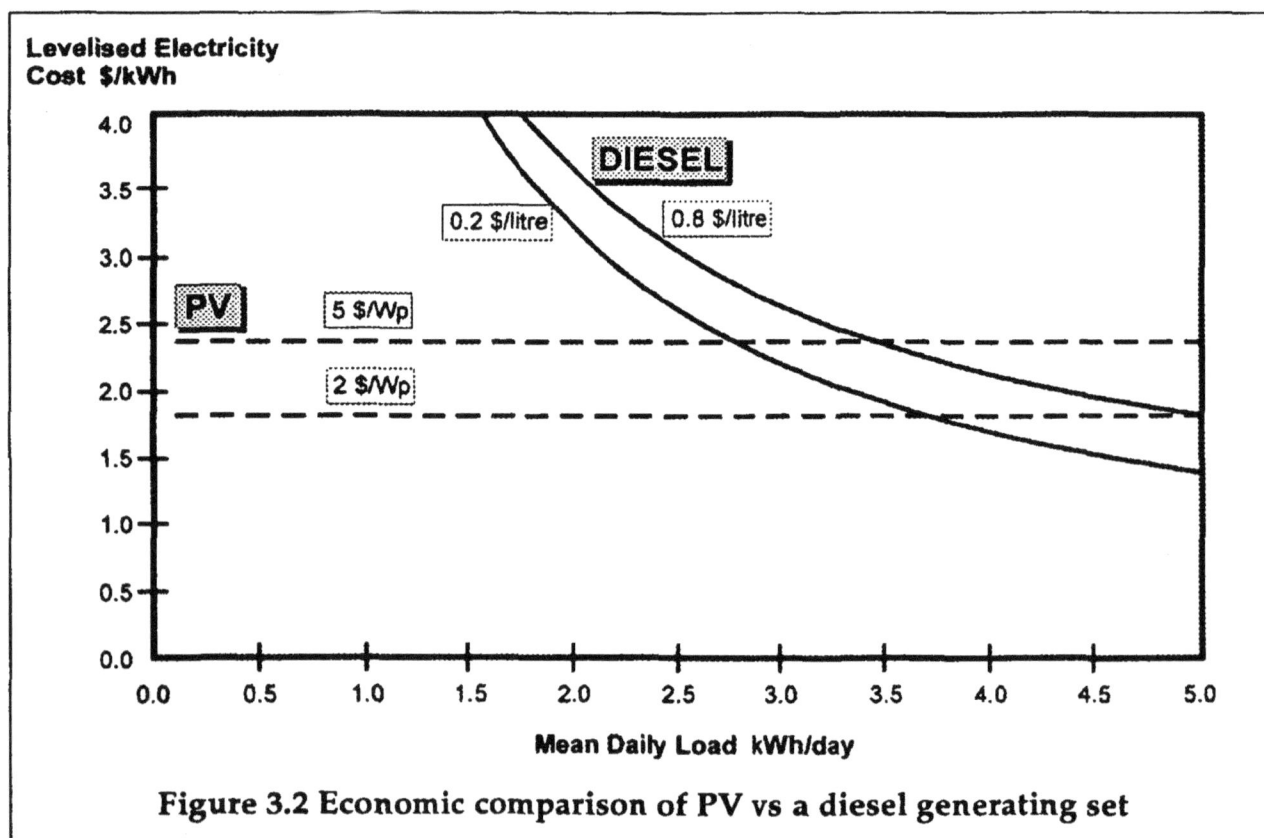

Figure 3.2 Economic comparison of PV vs a diesel generating set

PV vs Diesel

Figure 3.2 is a graph showing how the levelised electricity costs of PV and diesel options vary as the daily energy requirement increases from 0-5 kWh/day (using the same assumptions as in the life-cycle costing worksheets in Appendix 1). The effect of variations in both module price and fuel price are shown. At a module price of $5/Wp, the net cost per unit of electricity consumed is shown to be around $2.4/kWh, falling to $1.7/Wp if module price decreased to $2/Wp.

The two lines for PV are flat because the cost of energy from PV does not change with the size of the load: a PV system can be sized to meet any magnitude of load by adding the proportionate number of modules and batteries. Conversely a 5kVA diesel generator (the smallest size that is widely available) provides very expensive electricity when supplying small loads because the same start-up cost

has to be incurred whether the user needs 1kWh/day or 10kWh/day.

PV vs grid extension

Figure 3.3 compares the approximate levelised electricity costs for extending an 11kV grid over 1,5,10 and 20 kilometres to electrify a small village, compared with providing small PV installations for each household. The conclusion is that for small loads, it is not worth extending the grid by more than a few kilometres. PV is more cost-effective at small loads, but much more expensive at higher loads. The precise cross-over point between PV and grid extension is highly dependent on local circumstances.

3.3 Economics and real life

Although economic analyses have a key role to play in selecting energy technologies and

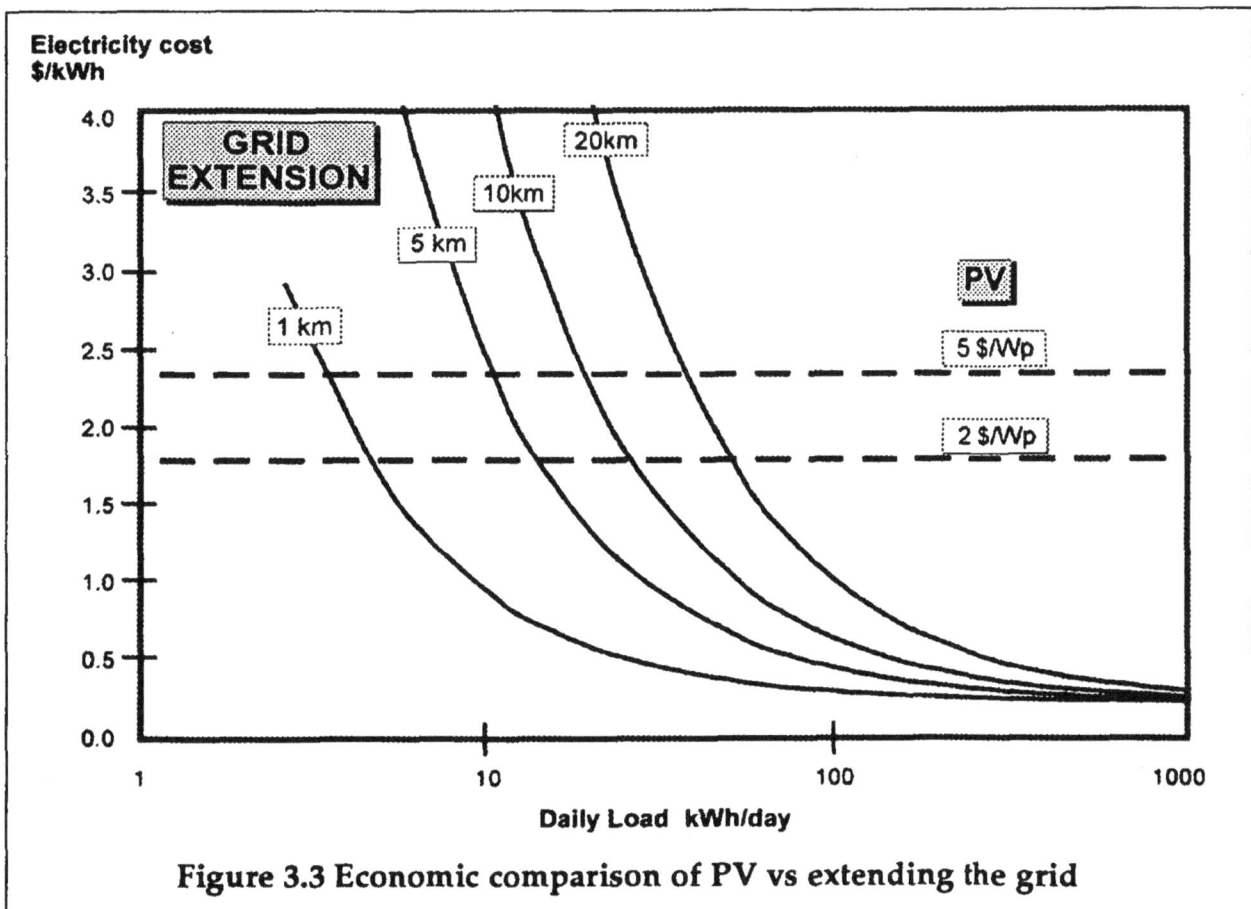

Figure 3.3 Economic comparison of PV vs extending the grid

planning projects, two further issues can affect the overall economic outlook:

- social, technical and personal factors can affect individual purchasing decisions

- governments can play a decisive role in distorting the economic situation

Real-life decisions

It has to be borne in mind that customers do not usually carry out a life-cycle cost analysis before deciding to make a purchase. Other influential factors are likely to be:

- the up-front cost

- the convenience and portability of the system

- the ease and speed with which the system can be installed and running

- the customer's confidence in the performance and reliability of the technology

- the maintenance implications and availability of spare parts

The customers' perception of these and other issues has a value which is not necessarily quantifiable in strict economic terms. A full comparison of the available options therefore becomes extremely difficult. Whatever the economics, a customer will not choose a diesel generator over PV if the requirement is for a system with as little maintenance as possible, nor will micro-hydro be chosen over diesel if power is wanted at short notice.

Financial or economic decision-making

Governments can strongly influence purchasing decisions because of their ability to either tax or subsidise in a variety of forms. Hence cheap goods in one country can be unaffordable in another despite identical manufacturing costs.

This raises the difference between taking a financial or an economic perspective. A financial analysis is carried out from the point of view of the private investor. Thus, the decision for a household or a company will be based on the real financial costs i.e. actual market prices, inclusive of local taxes, national taxes, import duties, and the cost of borrowing the capital to make the purchase. An economic analysis, however, considers projects from the point of view of the economy as a whole. It therefore looks at costs which exclude taxes and subsidies. An economic assessment is the most general case because it reflects the situation of a 'level playing field' in which the price of goods reflect their real costs.

Thus the economic approach must be interpreted correctly, or it can lead to misleading results if mistakenly viewed as a financial analysis i.e. from the buyer's perspective.

Continuous power:	1 W	10 W	100 W	1 kW	10 kW
Primary Cells	‖‖‖‖‖‖‖				
PV - battery	‖‖‖‖‖‖‖	‖‖‖‖‖‖‖	‖‖‖‖‖‖‖	‖‖‖‖‖‖‖	
Wind - battery	‖‖‖‖‖‖‖	‖‖‖‖‖‖‖	‖‖‖‖‖‖‖	‖‖‖‖‖‖‖	
Hydro-electric turbine				‖‖‖‖‖‖‖	‖‖‖‖‖‖‖
Diesel Gen. with battery			‖‖‖‖‖‖‖	‖‖‖‖	
Diesel Generator				‖‖‖‖‖‖‖	‖‖‖‖‖‖‖
Grid Extension				‖‖‖‖	‖‖‖‖‖‖‖

Figure 3.4 Typical economic power ranges for energy systems

For example, the local price of diesel fuel can be much greater than the imported price because of government taxation and high mark-ups in remote areas. In an economic cost comparison, rural electrification via diesel generators may look relatively attractive as a government project, but not for a private investor who may have to pay more than double the imported fuel price.

In carrying out a life-cycle costing exercise, the methodology is the same whether an economic or a financial analysis is being performed, however certain parameters will have different values. In general, a financial analysis will need to use actual market prices for the goods, a shorter period of analysis, and a higher discount rate. Hence a financial analysis usually leads to a higher annual cost for a particular scheme.

3.4 Summary of economic energy and supply options

Graphs of the types shown in Section 3.2 can be used to compare the full range of energy options against each other. In general, it becomes clear that each technology has a certain range of average energy demands which it is most cost-effective in supplying. Figure 3.4 indicates the appropriate size ranges for the most common options normally considered.

The broad picture can be described as follows:

- There will always be a market for portable throw-away batteries for the lowest levels of energy demand e.g. watches, torches and radios. Good rechargeable batteries with a PV battery-charger will prove cheaper than disposable batteries over a few years.

- For stand-alone domestic power supplies for lighting, TV, radio and fan, a battery-based supply charged by PV or wind is likely to be the most cost effective.

- For a number of domestic supplies e.g. for a small village, the economic choice lies between individual battery-based units or a centralised scheme with diesel generator or micro-hydro. For remote water-pumping, the choice is between PV, wind pump or diesel pump systems.

- For electrifying larger villages where 10's of kW are required, the choice is principally between micro-hydro, diesel, or extending the grid.

Dissemination Issues 4

There are many issues and barriers which are currently impeding the large-scale dissemination of renewable energy technologies (RETs). In developing countries, where RETs can have immediate effects upon living standards, it is even more critical that these issues are understood and overcome. These can be categorised as (i) institutional and non-technical issues and technical issues, such as training and (ii) technical infrastructure development. There are now many reports and papers which have been written on this subject. This chapter summarises these issues only and further information is available from references European Photovoltaic Industries Association (1996), Gregory (1993) and (1994) and IT Power (1996c).

4.1 Institutional and non-technical issues

Social and environmental awareness

Lack of knowledge, misunderstanding and bias have acted as deterrents to the wider utilisation of renewable energy systems. 'It doesn't work'; 'it's too expensive'; 'the technology is too complicated and the risk is too high for developing countries'; 'it's not appropriate' are just some of the arguments which have been used in the past. Whilst these were problems in the early years of development, they are no longer true. Many RET systems are technically mature and proven, particularly PV applications. Not only can they bring increased living standards to a community through the direct provision of electricity, hot

water and efficient cooking, they can also generate employment opportunities. Small businesses can be set up, such as sewing businesses and flour mills which use the electricity from renewable energy to drive their machine. Where there is adequate demand, businesses can be established to manufacture

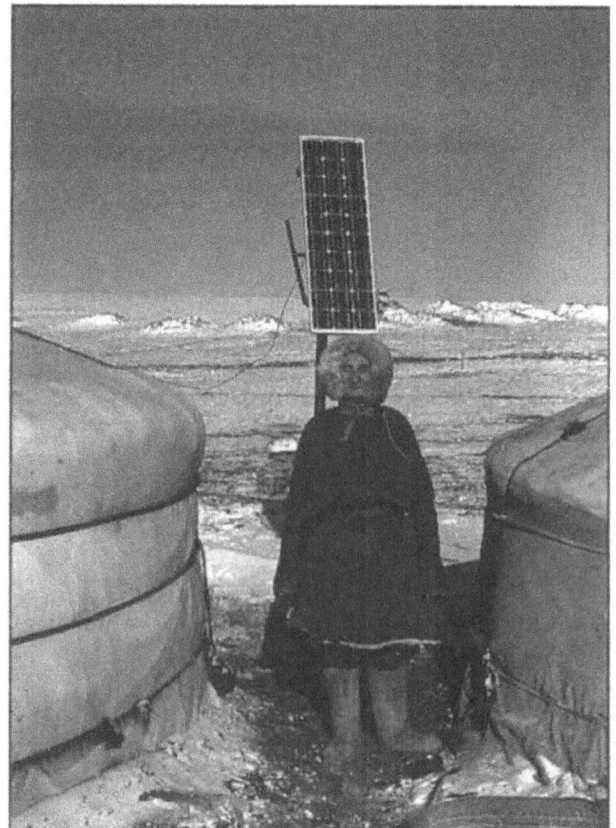

The modular nature of PV systems means that they can easily be taken down and reassembled Photo : IT Power

22

Solar water system, Mendefara hospital, Eritrea   Photo: Synopsis

components, such as wiring, solar water heaters, PV panels, solar cookers etc.

Direct social benefits of RETs installed in many rural areas of developing countries are access to lighting systems, water pumping, vaccine refrigeration, battery charging etc., all either previously unavailable or available only at high economic cost. Improved health of the population is possible through access to immunisation, clean drinking water and increased food production. Education can be enhanced through access to lighting for evening study and increased time for children to go to school by reducing requirements for fetching water and fuelwood.

There are more than a thousand health centres in developing countries which now use PV and solar water systems (WHO 1993). However, these are a fraction of the ultimate potential, despite the best intention of the World Health Organisation (WHO), United Nations Children's Fund (UNICEF) and charities such as Save the Children Fund. One of the impediments is lack of knowledge about the technologies themselves and funding mechanisms. This lack of knowledge can occur at local and institutional levels. The case study on the Democratic Republic of Congo describes how funding for health clinics can work, and how the systems can continue to be properly maintained once the donor organisation has withdrawn.

Without adequate information dissemination about RETs, their capabilities and the energy programmes themselves, a number of misconceptions may be perpetuated. These will lead to disappointment and collapse of the programmes.

User participation

The first issue which needs clarification right at the beginning of a programme is whether or not a community actually wants and is prepared to pay for the proposed system(s). It could be that lighting is not a priority, or water pumping for village needs is secondary to water for livestock or agriculture. Therefore, the first objective is to involve potential users and their communities to determine their real energy needs and to participate in the decision-making process of the programme. This includes technical decisions (what equipment to purchase etc.) and financial decisions (what repayments

---

Consequences of Top-down approach to RET implementation

In French Polynesia in the 1980s, a French aid programme installed a number of PV systems in remote villages. The purchase of these was subsidised by the utility, who also carried out the routine maintenance. However, at the end of 10 years a study carried out on the social aspects of this project found a general displeasure with PV. This was because there had been a general perception that the systems would provide electricity for far more services than was actually possible (lots of lights, TV, washing machines etc.). As a result, PV was seen as inferior to grid connection or the more abundant supply of electricity from generators (even though there was a high fuel cost) (Jourde, 1991).

23

they can afford, the type of financing scheme etc).

It is very important that the user does not have unreal expectations concerning the amount of energy which the PV or other RET system can supply. Otherwise, this can lead to dissatisfaction with the systems and, eventually, a high loan default rate.

Once the type, size and cost of systems have been established, and the type of fund decided upon, then the next issue relates to responsibility for the system(s). If the system is a community water pump, for example, but no-one in the community is responsible for its running, maintenance etc., then it is likely that the system will fail. For individual home or micro-enterprise systems, issues of maintenance and repair must also be worked out right at the beginning of the programme.

Financiers

Much work is needed to educate financiers, such as bankers, as to the merits of renewable energy systems. Typically, banks cannot afford to service large numbers of small clients, be they for micro-enterprise lending or renewable energy technologies. The reason for this is the high transaction cost per client and low returns to the bank. Many banks overcome this by charging high interest rates and/or bank charges. Banks, therefore, have little interest in lending for small-scale renewable energy systems, unless there are other opportunities presented by other interested organisations, such as NGOs or credit co-operatives. Lending through these schemes can be profitable for them, particularly if there is a savings component. Commercial and development banks can also act as a channel for international funds.

In chapters 5 and 7, the types of arrangements which can be made with banks and other financial institutions are discussed in more detail.

Many thousands of SHSs in Indonesia have been financed by a revolving fund.
Photo: IT Power

Government departments

One of the major challenges for governments of developing countries is to increase the standard of living for their burgeoning population. One of the ways to do this is to provide electricity to those people who are without it, as the provision of electricity is linked with economic and social development. Many people are increasingly forced to farm marginal land, and thus have a reduced ability to pay for services such as connection to the electricity grid. Many are leaving rural areas for the cities, putting unsustainable demands on urban areas and creating dismal living conditions for the migratory population. By providing sustainable and appropriate electricity in rural areas in the short term, rather than giving political promises of grid extension in some 20 years, the government can lower if not halt this migration. RETs, particularly PV and wind systems, can be a solution and should not be seen as a quick fix instead of grid electrification.

Energy is a cross-cutting issue, relevant to many ministries and disciplines, such as agriculture, resource planning, health and environment. However, officials in sectors outside energy often do not know about RETs, or do not know enough about them to specify project applications. Even within energy ministries in many countries, there is often a low level of knowledge on the technical advances and different applications. This once again can act as a serious deterrent to sustainable project development.

In many countries, national and/or regional governments have not yet made RETs an integral part of their energy strategies. Whilst this applies to developed as well as developing countries, for the latter a decision of integration is critical for the following reasons:

- renewable energy technologies do not require expensive imported fuel, paid for in valuable foreign exchange

- local manufacture and associated industries for RETs can bring employment to rural areas, thus increasing the flow of wealth

- RETs can be used in many different sectors, and can be less expensive than alternatives (see chapter 3)

However, recently, general awareness about renewable energy technologies and their appropriateness for electricity production in remote areas in particular has increased in many governments. The challenge for renewable energy funding is now to tap into funds for rural development, agriculture, healthcare, micro-enterprise development etc., so that they are mainstream activities, not fringe Research and Development projects.

Governments can also facilitate the installation of good quality installations by funding Centres of Excellence for renewable energy technologies. Such a Centre could have a technical resource sufficient to undertake local testing and quality control of equipment. It could also undertake training activities for installers and technicians, and conduct information dissemination campaigns to the general public. In India, the Indian Renewable Energy Development Agency (IREDA) also handles the dispersement of international loans to renewable energy projects (Bakthavathsalam, 1995).

Governments can also assist the market to develop by providing opportunities for demonstration of RETs systems. In the case of Indonesia, the government commissioned the PV electrification of Sukatani village. This provided an excellent demonstration for other villagers, bankers, civil servants etc. to see working. It has also been catalytic in Indonesia's commitment to PV electrification in rural areas (see Indonesia Case Study for more details).

Utilities

The traditional development model for electricity utilities is to electrify the areas with the highest number of customers first, and thereby receive the highest return on investment. This means that rural areas do not usually have access to a reliable source of energy if the country is poor or has a sparse rural population.

As electricity organisations/ utilities grow, they tend to become more conservative and less likely to fund methods alternative to traditional generation and grid distribution, because this could decrease their profit mar-

In 1996, the South African utility, Eskom, installed 1100 PV systems in rural schools. This is part of the government's reconstruction and development programme   Photo: Eskom

gins. Only a change in corporate attitude, coupled with other stimuli (such as seeing RETs as a mechanism to increase market share and profitability, or through direct government directives) have managed to change this stereotypical attitude world-wide.

There are, however, some direct examples of positive action towards PV which have been taken by utilities, for both grid-connected and stand-alone systems in developed and developing countries. For instance, in South Africa the government of National Unity directed Eskom to electrify 15,000 schools and health clinics in remote areas with PV. Although Eskom has experienced problems, for example sourcing adequate financing for this development, many installations have already been realised.

Utilities can participate in rural electrification using RETs, particularly PV, in a number of ways. They can be a co-partner in a scheme to bring renewable energy systems to unelectrified areas. They can operate as Energy Service Companies (see chapter 4), owning and operating PV equipment themselves, or otherwise facilitating its installation and operation (such as Southern California Edison and the Sacramento Municipal Utility District).

4.2 Technical issues

Technical Infrastructure Development

Accessible after-sales service is important to the continued running of any fund. Spare parts must be available at reasonable costs and relatively quickly. The density of systems in an area will influence the provision of services in that area. For instance, it has been estimated that a density of 200 systems is necessary to support a trained technician and adequate supply of equipment (IT Power and Eurosolar, 1996).

Example Commercial Barrier

An example of how the lack of commercial activity can be a barrier to the use of RETs can be found in Mongolia. More than 80,000 families in Mongolia are nomadic livestock herders, and thus grid-connection is not possible for them. As a result of demonstration programmes supported by the national government, UNDP and Japan, there are now more than 3,000 small PV systems in use in the countryside. Systems range from 10 to 200 Wp, and are used to provide electricity for lighting, TV and radios. User acceptance is good as the systems supply basic electricity requirements, are modular, and are easily transportable.

Yet a 1994 mission to the Gobi region found little evidence of commercial activity, and no infrastructure set up to provide for the procurement of new systems, spare parts, repairs or even replacement batteries. The principal source of batteries and other system components is Ulaanbaatar, which can be several hundred kilometres from the users of the PV systems. This has resulted in some systems performing poorly because they are using batteries which are at the end of their useful life (Chadraa et al, 1994).

Natavada Health Centre: making sure that the orientation and tilt angle are correct. Photo: IT Power

26

PV pumping systems provide potable water close to the point of use, Banyankakou, Mali   Photo: IT Power

any parts including the batteries, and collection of repayments) right at the beginning have been the most successful.

The best way to train the user is during installation, with retraining if necessary carried out during routine maintenance by the technician. Simple measures, such as the provision of a user's manual as well as a poster permanently fixed onto a nearby wall or directly onto the system, can be used to reinforce important messages.

Standards for components and quality control

PV equipment which needs to be available includes: batteries, wiring, modules, regulators and lights. Depending on the maturity and size of the local market, some of these components could be manufactured locally.

Dissemination and training

Training of local installers and maintenance personnel is important to the continued reliability of the systems. Training programmes for local installers and maintenance technicians need to be integral with any programme. It is also important to define at the outset of a financing programme whose responsibility these training activities are: the fund, the private sector, the distributor or the supplier. This responsibility can be undertaken directly by the fund, or subcontracted. To date, PV financing schemes have adopted different approaches, depending on local conditions. These schemes have varied in their effectiveness, but those which have defined the responsibilities for training (and other issues such as maintenance of the systems, replacement of

There are many very good and reliable RET system components on the international market. PV systems can have warranties on components of up to 20 years. The majority of modules have been tested for quality, either through a national laboratory in their own country or at an international test centre, such as the European Union PV Test Centre at Ispra, Italy. They will be labelled with this on the rear of the panel. Other components must also be checked for quality before being purchased.

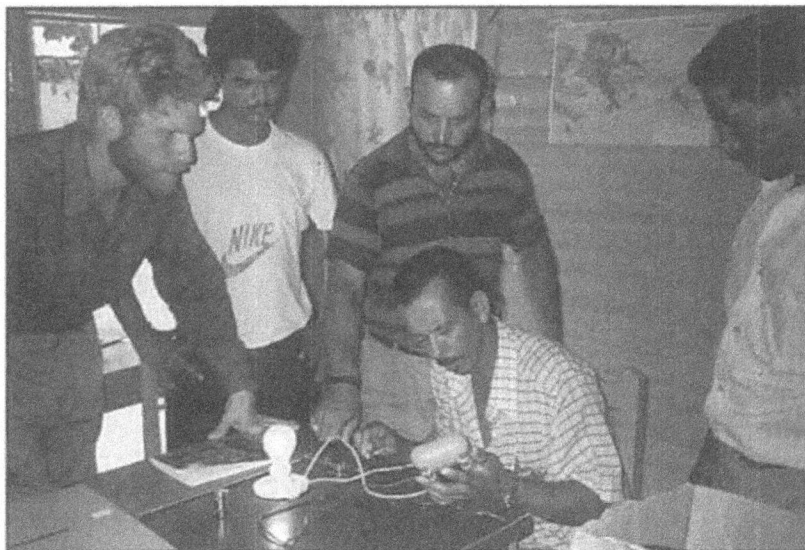

Technician training, Bella Vista, Dominican Republic.   Photo: Enersol

Maintenance requirements for each main component within a typical PV system:

- *The PV array* needs only regular cleaning with a damp cloth once every two to three months. In especially dusty regions or during dry season months, cleaning may need to be more frequent. A periodic visual check for terminal corrosion and loose wiring is advisable as well as making sure that growing trees do not block the sunlight from the modules. These actions are usually sufficient to maintain the PV array electrical output.

- *The Power Conditioning Unit (PCU)* includes the regulator, inverter and maybe a maximum power point tracker. These do not need much maintenance, only regular routine checks by qualified electricians.

- *Most types of batteries* (i.e. vented types) need regular and essential maintenance. This consists of topping them up with distilled water, checking and removing any build up of corrosion around the terminals, and checking that the casing is sound without cracks or leakage.

- *Cabling components* do not require regular maintenance but rather visual checks to detect any faults such as cables 'eaten' by rodents.

- *End-use equipment*, in some cases, will be models specific for photovoltaic use (e.g. vaccine refrigerators, high efficiency water pumps). In such instances, additional maintenance tasks may be necessary to ensure satisfactory performance. These are easily undertaken by a qualified technician.

There are products on the market which are not reliable, and, when they fail, give RETs a bad reputation. For instance, if a PV module is said to be able to produce more electricity than is possible, then the consumer will be very disappointed and feel cheated when this does not happen. Similarly, if a battery is said to last 3 years, and it lasts only a year, then dissatisfaction with the system will occur.

There are various ways of overcoming this. One is to require certification on all products. Another is to require that proper specification is given for the product from the manufacturer or supplier, both on attached labelling and packaging. The International Standards Organisation (ISO), World Bank, PV industry and other organisations are working on a Global Approval Programme (GAP) for PV systems.

Installation, operation and maintenance

All RET systems require installation and some maintenance. PV systems have, in the past, been 'over-sold' as requiring no maintenance. Although it is true that minimal maintenance is required, it is also true that this maintenance is vital. In general, the overall maintenance required by a PV-powered system is much less than most similar systems powered by conventional sources such as diesel or petrol generators.

Provision for maintenance is necessary in the programme, though, like training activities, this can be the responsibility of the fund or the local private sectors. Maintenance and repair services can be supplied and paid for by the fund, with the cost factored into the financed price of the system. Batteries need to be replaced every 3 to 5 years, and the responsibility for paying for this cost also has to be clear from the outset of the programme. Chapter 5 discusses these responsibilities and options in more detail.

All systems (except for portable lamps) require proper installation. Those which have not been installed correctly are likely to operate unsatisfactorily. However, installation of PV systems, for example, does not necessarily represent any particular or major problems, especially when they are sold in 'kits'. This is provided that the kit contains all necessary components, from the module to the cable fixing components, along with instructions in the appropriate language. In the case of well designed 'plug-in' domestic

28

lighting kits, the installation can be completed by the owner. For community installations, such as for a hospital or school, the installation of PV systems should be undertaken by qualified technicians with the same attention to correct installation as for conventional power supply systems. This, in turn, will lead to satisfactory long term results. In some programmes (e.g. GEF PV programme in Zimbabwe) installers are accredited to the programme and have to maintain satisfactory quality in order to continue as a qualified installer for the programme.

4.3 Conclusions

There are many issues which affect the success of renewable energy programmes in developing countries. Of these, probably the easiest to remove for SHSs are technical barriers, as the industry is mature and many products have an enforceable guarantee, and there are thousands of examples of working systems around the world. Non-technical barriers are more difficult to remove, as often they are caused by political agendas. However, many non-technical barriers are also very easy to remove, such as lack of information and inadequate local participation in decision-making. No barriers are impossible to overcome.

The next chapter continues looking at overcoming barriers, by presenting the case for different types of financing schemes at a local level.

Role of Institutions 5

Projects for financing renewable energy investments are influenced by the type of institutions and organisations in the country. These affect all areas of the project: financial aspects (such as funding and access to credit), technical aspects (for example through the provision of technical assistance), administration, and the long term achievement of goals. Different types of organisations are usually responsible for different aspects, and their competence to undertake the tasks required by a project is vital to the success of the project. Choosing the wrong type of organisation to administer a project or loan fund can spell disaster. However, if the right choices are made and infrastructure investments made (either private or public) to strengthen these so that they can carry out the activities required by the project, then there is more likelihood that the project will succeed.

One of the major challenging tasks in financing and marketing RETs in the developing world is to build efficient structures for local financial intermediaries. It is possible to do this by combining the resources of the public sector and the creativity and enthusiasm of those at the grass-roots level. There are a number of alternatives for sourcing funds at an international level, and dispersing them at national, regional and local levels.

This chapter looks at the main institutions and types of organisations that can impact upon RETs financing programmes in developing countries.

5.1 International institutions

International institutions have an important role to play in shaping the institutional and organisational reforms needed to implement renewable energy schemes in developing countries.

For developing countries, donors such as the World Bank and the development agencies have acted as catalysts for transfer of renewable energy technologies. They are an important source for project funding, indigenous training and infrastructure development. These institutions have increased the flow of information about PV from industrialised to developing countries. They have aided in the development of sector programmes and projects, assisted in project execution (financing, technical assistance, training, etc.), and carried out post-project appraisals. The extent to which RETs have been included in these activities has varied enormously from institution to institution, and even within departments in individual institutions.

Multilateral Agencies

The World Bank Group

The World Bank Group consists of four organisations, all of which have the capacity to finance renewable energy projects. However, only the International Bank for Reconstruction and Development, IBRD, has done so in the past. The reasons why the others have not done so vary, but the major reason is that renewable energy projects were not seen as either a priority or cost-effective solution. Both these viewpoints are currently changing, though the size of the impact which this will have for financing renewable energy technologies is unknown at present.

The International Finance Corporation (IFC) promotes development by supporting private sector in developing countries, and helping to mobilise domestic and foreign capital for this purpose (Environment Matters, 1996). The IFC makes long-term loans, including subscriptions to share capital of privately owned companies. However, it cannot hold more than 35% equity in a concern, nor be the largest shareholder. It is becoming increasingly interested in lending for renewable energy programmes, provided that the lending criteria are met. However, although they commenced establishing a fund for renewables in 1995, it will take some time before it is possible for developing countries to access funds through this mechanism. This lag is due to the administrative procedures of the IFC. Similarly, World Bank projects can take over five years from project submission to realisation and the release of funds.

The International Development Association (IDA) lends on favourable terms to the least developed countries, i.e. those unable to access IBRD loans. Although lending for renewable energy programmes is not precluded, no such funding has occurred to date.

The Multilateral Investment Guarantee Agency (MIGA) helps developing countries attract productive private and public foreign investment. Its facilities include guarantees against non-commercial risk, and a programme of consultative and advisory services to improve the environment for foreign investment in member countries. It has not supported renewable energy projects to date, but, if the right criteria are met, it is not precluded from doing so in the future.

The International Bank for Reconstruction & Development (IBRD, or more commonly known as the World Bank) borrows the bulk of its money from international capital markets. It then lends to credit-worthy developing countries for projects that promise real economic return for the country, on more favourable terms than they could otherwise

obtain. It has supported PV and other renewable energy technology projects (e.g. wind, micro-hydro) for many years. However, these projects did not form part of a renewable energy strategy. Rather they were in response to requests from individual countries.

The World Bank is an implementing agency for the Global Environment Facility (see section on GEF) and the Montreal Protocol (MFMP). The MFMP funds projects which assist developing countries to comply with objectives for reducing emissions and production of ozone destroying substances.

There are now a number of renewable energy programmes within the World Bank. These include:

The Solar Initiative The aim of the Solar Initiative, launched in March 1994, is to hasten the commercialisation and employment of RETs in developing countries. This initiative has the ability to tap Global Environment Facility (GEF) funding resources for renewable energy projects. The Solar Initiative also aims to be an overall co-ordinating body, assisting with strategic decisions for renewables and catalysing accelerated research, development and demonstration of commercial and near commercial renewable energy technologies. PV projects in developing countries are of particular importance within its strategic programme, and there are currently over 40 projects in various stages of preparation.

The PV Market Transformation Initiative The goal of the PV Market Transformation Initiative is to accelerate the commercialisation, market penetration and financial viability of PV technology in developing countries. It also promotes the large-scale use of PV as one of the best long-term prospects for a low carbon energy future. The project will be completed in two phases; the appraisal phase commenced in May 1997 to demonstrate the viability of the proposed initiative, and to design the optimum framework for the subsequent implementation phase. The implementation phase will disburse approximately $25 million from the GEF to provide direct investments of between $1 and $5 million in several innovative, private sector projects. The

FINANCING RENEWABLE ENERGY PROJECTS

PV-powered pump and security lighting, Swaziland   Photo: IT Power

disbursement terms will be finalised during the appraisal phase, but funds will be disbursed in three countries, namely India, Kenya and Morocco. The minimum leverage of private capital has been set at 3:1, and it is anticipated that this will be sufficient to catalyse the market, stimulating the formation of joint ventures, pioneering innovative financing schemes and promoting public-private partnerships. Subject to final approval of the programme, PVMTI implementation will start towards the end of 1997. (IT Power, 1997; IFC/GEF, 1996).

Asia Alternative Energy Unit (ASTAE) ASTAE was created as an initiative of the ASEAN FINESSE programme, and since has assisted in identifying opportunities for renewable energy programmes in developing countries in the Asian region. ASTAE's role to date in assisting renewable energy projects to develop has included the funding of country studies and workshops (Schaeffer, 1993).

Energy Sector Management Assistance Programme (ESMAP) Supported by the World Bank, the UNDP and other United Nations and bilateral organisations, ESMAP identifies and analyses the most serious energy problems in developing countries, and proposes priority investments and technical assistance projects to address these energy problems.

Financing Energy Services for Small-Scale Energy Users (FINESSE) was launched by

the World Bank with the US Department of Energy and the government of The Netherlands in 1989, to promote the market implementation of renewable energy technologies. It aims to facilitate the identification of renewable energy potentials within participant countries, and to assist them to establish priorities for utilising these sources. It will assist embryo industries to exploit resources, generate wealth within the country and increase living standards. To achieve its aims, FINESSE identifies and supports the financing of small-scale energy users' projects by channelling low-interest loans and grants through a range of private, utility, non-governmental and commercial lending intermediaries.

The first FINESSE programme was in Asia, with Indonesia, Malaysia, the Philippines and Thailand participating. In 1994 a FINESSE programme for South America commenced, and in 1996 a FINESSE programme for Southern Africa was launched. This is being executed by the United Nations Development Programme (UNDP) and the Southern Africa Development Community (SADC) (UNDP, 1996).

Global Environment Facility (GEF) was initiated in 1993 as one of the major recommendations of the United Nations Conference on Environment and Development, and was established on a permanent footing in March 1994. It is a financial mechanism that provides grants and concessional financing to developing countries for projects and programmes aimed at protecting the global environment.

The GEF priority areas are:
- protection of biodiversity
- reduction of greenhouse gas emissions
- protection of international waters
- reduction of the ozone layer depletion

The initial funding for the GEF in its pilot phase (1991-94) was around $800 million,

Large scale PV grid connected system, co-financed by IREDA, Tamil Nadu
Photo: IT Power India

domestic public and private, multilateral, bilateral and NGO sources.

There have been a number of large RETs projects funded by the GEF (Kozloff, 1994). PV is currently the number one renewable energy technology priority area for the GEF. The largest PV projects funded to date are the Renewable Resource Management project in India and the PV for Household and Community Use project in Zimbabwe. GEF funding for this type of project is accessed through a national government process in each country (GEF, 1992a; GEF, 1992b).

with which it supported 113 projects. In 1995, after a reconstruction process, the 87 funding countries reaffirmed their support of the GEF and pledged $2 billion for its next phase.

The GEF is jointly managed by the World Bank, United Nations Environment Programme (UNEP) and the United Nations Development Programme (UNDP). The World Bank is responsible for investment projects and to act as a repository for the Global Environment Trust Fund. UNDP is responsible for technical assistance and training activities. It also helps to identify and prepare projects through a pre-investment facility. It runs the Small Grants Programme. UNEP provides environmental expertise to the GEF process, and ensures that GEF policy and projects are consistent with existing environmental treaties. The GEF also has access to the Scientific and Technical Advisory Panel (STAP), which provides appropriate advice to the Council (World Bank, 1993; Environment Matters, 1996).

The GEF provides money for demonstration, pilot and long term projects in developing countries. In effect, it 'buys down' environmental technologies that are close to being commercially competitive with existing polluting technologies, but which are still more expensive. The difference between the GEF grant and the total project cost is financed by

The GEF Small Grants Programme provides funding for projects that demonstrate community-based strategies and technologies. In the past, this was administered by a centralised body meeting in New York. In 1996, this process was decentralised, and an advisory committee established in each country

> ## The Indian Renewable Resource Management project
>
> The Indian Renewable Resource Management project promotes and commercialises investments in wind farms, small hydro and PV systems, through the provision of below-market loans to system investors, who are primarily from the private sector. There is a public education programme to popularise the technologies and to explain their function and capacity. The project is being implemented by the Indian Renewable Energy Development Agency. The overall project cost is $195 million, with the GEF component being $26 million. Co-financiers include the IDA, Danish Development Agency (DANIDA) and the Swiss Development Co-operation (SDC) (GEF, 1992a; Environment Matters, 1996; Bakthavatsalam, 1995).

33

for the assessment of small grants. NGOs are able to apply for funding under the Small Grants Programme, and a Guide for NGOs to the GEF has been published (CNE, 1996).

There have been a number of small grants given for projects and dissemination activities; for PV financing via revolving funds (e.g. ADESOL, Dominican Republic); and for studies to facilitate future projects.

Agencies of the United Nations

The majority of agencies of the United Nations have no impact upon renewable energy technologies. There are numerous reasons for this, including: the inappropriateness or irrelevance of such technologies for their activities, the agency is not directly self-financed from the Economic and Social Council (ECOSOC) or the agency's policies and programmes are dictated by other organisations.

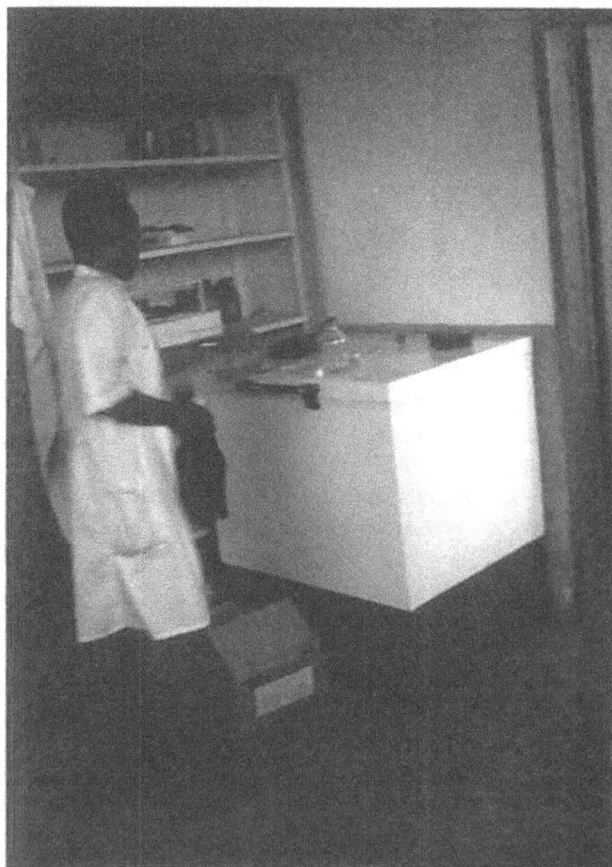

PV powered medical refrigerator for storing vaccines, blood and for ice-making   Photo: IT Power

However, there are a few agencies that incorporate PV within their activities, albeit on a small scale. In doing so, they assist in promulgating PV. Co-operation with their programmes can be important for the large-scale dissemination of renewable energy technologies.

United Nations Industrial Development Organisation (UNIDO): is not directly funded from ECOSOC, but is funded through other agencies, such as UNDP. It is also supported by countries, though the USA is ceasing its contributions in 1996, which will significantly reduce its already meagre RETs funding resources.

UNIDO does not fund projects, and has not undertaken any significant renewable energy projects. It has facilitated a number of small renewable energy country studies, facilitated workshops, and assisted in the development of Centres of Excellence in PV, such as the Centre for Excellence (CASE) in Perth, Western Australia. CASE is funded primarily by the Australian Federal Government and the State Government of Western Australia (Bromley, 1995).

United Nations Children's Fund (UNICEF): does not have any RETs programmes. However, through its country programmes (which are formulated in consultation with government officials), it has purchased over 5,000 PV vaccine refrigerators, together with lighting kits, radios, water pumps, etc. These are installed in health clinics and hospitals in remote areas of the developing world (WHO, 1993).

World Health Organisation (WHO): does not have a renewable energy programme as such. However, it has encouraged the installation of PV vaccine refrigerators in developing countries (as well as the use of other solar technologies in health programmes), and has provided small amounts of funding for technical assistance to a number of these programmes. WHO has participated, though the Expanded Programme on Immunisation, in other organisations' RETs programmes, such as the Health Programme in the Democratic Republic of Congo, which was primarily funded by the European Development Fund, 1988 - 1991. From 1984 to 1991, this programme installed over 750 lighting kits and 100 vaccine refrigerators in rural health centres in five regions in the Democratic Republic of Congo. WHO has also funded many studies into the use of biomass in developing countries, and its impact on women's health and well-being.

WHO has developed a Standard Performance Specification & Test procedure for PV vaccine refrigerators, and developed information sheets from the results of these tests. Organisations installing vaccine refrigerators, such as UNICEF, often specify in their call for tenders that the refrigerators must be accredited to or meet WHO specifications.

WHO has also collaborated with research institutions and manufacturers to develop new products, improve reliability, reduce costs of existing equipment, and carry out long-term field testing. It has a course for technicians which it runs biannually in Asia (Zaffran, 1991).

United Nations Conference on Trade and Development (UNCTAD): The principal function of UNCTAD is to promote international trade, particularly between countries at different stages of development. Its priority is to assist Least Developed Countries (LDCs). In the early 1990s, its influence and funding were reduced. However, in late 1995, there was a renewed commitment by the non-aligned countries to increase its activities and funding (Go Between, 96/97). It is not known how this will affect its ability to fund or participate in future RETs projects and programmes. To date it has only funded a few very small studies on renewables.

United Nations Education and Scientific Organisation (UNESCO): UNESCO has provided funding for a number of networks, courses and books on solar energy. It is also co-ordinating the World Solar Summit process, which kicked off with the Solar Summit in Paris, June 1993 and was officially augmented by Heads of State in 1996. The Harare Declaration ratified the Solar Decade (1996-2005), and identified 300 solar projects for implementation programmes (Berkovski, 1995). UNESCO does not procure equipment or participate in-country.

Economic Commissions

There are five Regional Commissions: the Economic Commission for Africa (ECA), Economic and Social Commission for Asia and the Pacific (ESCAP), Economic Commission for Latin America and the Caribbean (ELAC), Economic Commission for Western Asia (ESWA) and the Economic Commission for Europe (ECE).

The Regional Economic Commissions do not fund projects. Instead they assist established institutions to attract funding for regional projects. These, in turn, supply development aid. They mobilise funds from affiliated UN agencies and funds, the use and transfer of which are then governed by the regulations of the donor agency.

The Commissions thus do not procure renewable energy equipment or run large renewable energy programmes. The possibility to do so is limited by their funding and the priorities of member countries. They have all funded small renewable energy projects, such as seminars and workshops on renewables, networks of experts, training courses, publications, studies and small demonstration projects. No large or significant projects have been funded by the Economic Commissions in any of the regions.

Regional development banks

There are six regional development banks: the Caribbean Development Bank, African Development Bank, Inter-American Development Bank, European Bank for Reconstruction and Development, Asian Development Bank and the Inter-American Development Bank. Although renewable energy and energy conservation often appear in the energy policies of the regional development banks, none have an integrated strategy to encourage the inclusion of renewable energy technologies, either within their overall lending programmes or within sectors. All of the development banks have funded renewable energy projects in the past, but these have been small and funding has not been not sustained.

However, they are well placed to support significant RETs projects, and the Asian

Development Bank in particular is looking at supporting PV and biomass projects in developing countries.

Bilateral Agencies

Bilateral agencies provide millions of dollars each year to countries less well off than their own. There are a number of bilateral donors who have included renewable energy technologies in their programmes and projects. The success of these projects has varied enormously from country to country. Influences include the suitability of the technology, the design of the programme, the level of funding, and the involvement of part or all of the project actors and recipients.

The European Development Fund (EDF) has initiated and funded the world's largest developing country PV project. It was a component of their Regional Programme to Combat Desertification in the Sahel (CILSS), in West Africa. This commenced in 1993, and followed on from the success of the first PV pumping project in Mali (which started in 1977).

Around $35 million was allocated for the procurement of 829 PV pumping systems, 530 community lighting or refrigeration systems, and related equipment for water supply for nine Sahel countries. These installations were completed in 1995 (Makukatin et al, 1994; EPIA, 1996).

A well equipped with a PV pump can provide potable water for many nearby villages
Photo: IT Power

The European Commission has a number of programmes which co-fund projects in developing countries. These include programmes within the Directorate for Energy (DG XVII) and the Directorate for Research and Development (DG XII). The Directorates for Development (DGs I and VIII) are funded by the EDF. The Power for the World initiative has been funded by DG XII. This initiative, which was launched in 1993, recognises that there are over 2 billion people in the world without electricity and any likelihood of receiving grid electricity in the near future. It proposes PV systems as the most economical means of providing these people (primarily in rural areas of developing countries) with electricity, and sets out ways in which this can be achieved (IT Power and Eurosolar, 1996).

The German Agency for Technical Co-operation (GTZ) has undertaken many PV projects in developing countries, since 1979. These include initiatives in Algeria, Tunisia, Morocco and the Philippines. In Burkina Faso, GTZ was responsible for the installation of 48 PV lighting systems. The continuing maintenance of these systems is carried out by local technicians (Biermann et al, 1995). However, a change of policy has meant that GTZ is now scaling-down its commitment to this type of project.

The Danish Agency for International Co-operation, (DANIDA) has not undertaken any major programmes supporting the market development and application of PV. However, it has supported PV pumping activities in India, and technology transfer of balance of systems components in Zimbabwe.

The United States Agency for International Development (USAID) has funded a number of PV-related projects. These include the PV pumping project in Morocco (which included the training of installers and maintenance technicians), and financing a market study that led to the facilitation of the financial package for a Sri Lankan PV manufacturing and marketing company. In 1984, a PV rural electrification project was set up in the Dominican Republic, using USAID seed money to install between 30 Wp and 50 Wp PV lighting systems. The income from charging for these systems has allowed further equipment to be purchased and the project is now self-financing, with more than 1,500 systems installed. USAID has also facilitated a number of training and demonstration schemes in many other developing countries. It is now active in conjunction with other US organisations, such as the Export Council for Renewable Energy (ECRE) and the Institute For Renewable Energy and Energy Efficiency (IFREE) in promoting the development of PV in Southern Africa and Latin America.

The Agency for Development Co-operation in the Netherlands (DGIS) is a significant

Training installers is a critical element of any project (Mali). This can be funded by the project or by the private sector. Photo: IT Power

supporter in developing financing mechanisms for renewable energy. Activities supported by DGIS include the creation of the India Renewable Energy Development Agency (IREDA), co-financing of PV projects in Indonesia, co-financing of the FINESSE activities in Asia and the 1996/97 FINESSE projects in the SADC countries of Africa.

The Swedish International Development Agency (IDA) has supported small and large scale energy projects for many years. The Stockholm Environment Institute (SEI) has facilitated a number of projects on rural electrification, improvement of traditional small-scale industries utilising biomass, gasification projects, etc. In 1996, IDA developed a new energy policy with five major priority areas:

- legislation and regulatory systems
- institutional capacity building
- energy efficiency
- use of renewable energy
- support to specific groups such as women

The new policy builds upon the principles of sustainable development and the recognition that development assistance can play a catalytic role in applying these principles.

5.2 National institutions

Governments

Proactive government policies and actions are vital for the successful implementation of renewable energy schemes in developing countries.

At present, hidden and direct subsidies are given to fossil fuels. For example, many developing countries have subsidised kerosene prices, and most developed countries have indirectly subsidised electricity prices, through mechanisms such as life-cycle cost accounting. This means that there is no incentive for the private or public sectors to try and spend more money on energy efficiency or alternative energy sources if more convenient and cheaper fuels are available. Proper pricing is a pre-requisite for optimum allocation of resources, leading to the worldwide cost-effective development and utilisation of RETs.

Government policies on import taxes for renewable energy systems can greatly help or hinder the affordability of the system. For example, the existing import tax in the Dominican Republic is set at a rate of 70% or more on every commercially imported good. An import tax break, or tax reduction on PV panels and related materials would reduce PV costs substantially and assist market development in the country.

Governments can work with international lending institutions to establish the credit channels necessary to provide rural households with the initial capital needed to buy the PV systems. In many countries, it is not possible to establish these channels without government backing, and indeed, in many countries NGOs are not allowed to act as credit and savings organisations. If this impediment is removed, and adequate regulations are put in place to ensure sound financial management, then more appropriate financing mechanisms can operate at a local level.

Technical regulations for PV systems in developing countries tend to be less stringent than those found in the developed world. However, it should only be a matter of time before these countries enforce tougher regulations, and insist that quality products are sold in the market.

One way to oversee the quality of systems used in a country is the establishment of regional or national Centres of Excellence. These Centres could be responsible for:

- training installers, technicians and scientists

- co-ordinating / undertaking product certification

- R&D in renewable energy technologies

- dissemination campaigns to potential users

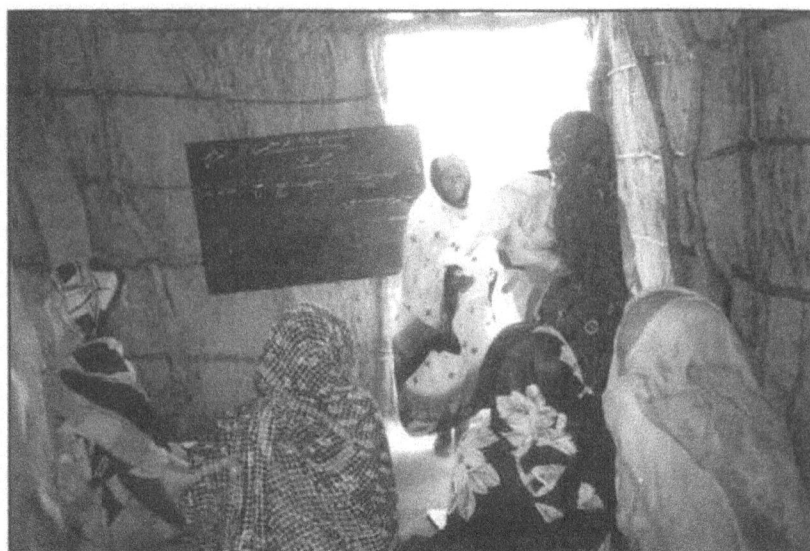

In El Ga'a, the Sudanese government and the UNDP are supporting this women's micro-enterprise centre. The PV lighting enables women to make haberdashery articles to sell Photo: IT Power

These activities need to be funded at a level which will ensure long term stability, retention of qualified staff, international recognition (which in turn can attract international funding), intellectual and skills wealth creation and the ability to make sure that RETs projects within the country have quality products and are properly administered (Pereira et al, 1995).

Some governments of developing countries are actively encouraging the development of renewable energy markets. For example, India, Zimbabwe and Brazil have large GEF-funded programmes. The Government of Indonesia (GOI) has set a target of 50MWp for PV rural electrification of 1 million households by 2005. The World Bank / GEF are providing 200,000 SHSs to assist the GOI in attaining this target. These are all technology-specific programmes, funded specifically from international sources.

In Mexico, the poverty alleviation programme of the national government contained a component for PV rural electrification. This programme did not receive international funds, was not specifically to develop the market and did not have a major cost-recovery component. However, it was very successful in acting as a catalyst for the development of the PV market in rural areas of the country.

PRONASOL Programme

The Mexican PRONASOL (National Poverty Alleviation and Development) developed infrastructures in less developed regions and communities, and ran from 1988 to 1995. It included rural electrification programmes of the utility, and a special budget to electrify communities not included in short term plans for grid extension. Installation of individual PV home systems was undertaken in the smallest, most dispersed communities and those farthest away from the electric grid, with no hope of being connected to the grid for at least five years.

There were many stages to the PV electrification programme. Firstly, suitable communities were identified, and PV promoted to them. On commitment to the programme, each community organised a local electrification committee, and promised to look after the system(s). A formal petition was filed with their local government, and all petitions from a single state were integrated into a portfolio, which the state government submitted to the federal government for funding.

Approved petitions were developed into engineering projects, normally by private companies under contract to the government. The utility, Comision Federal de Electricidad (CFE), undertook review and gave technical approval. Approved projects were let for tender by CFE as a complete package of design and installation, plus user training and participation, and infrastructure development for maintenance, and installation. This process assisted in maintaining quality and lowest prices.

There were two categories for PRONASOL project financing.

- *Productive users* included agro-industries and similar applications, and were required to show economic use. Funding was by a preferential loan scheme, developed by the Mexican development banks, and money loaned had to be repaid from the projects' economic surplus.

- *Life improvement* projects, such as electrification of individual houses (lighting and some radio and TV), street lighting, lighting for public buildings (schools, communal houses, health clinics etc.), educational TV, telephone, vaccine refrigeration and water pumping for human consumption. These projects were supported by government grants: 50% of total project costs from the federal government, 30% from the state governments and 20% from a combined effort between the local government and the community (or individuals, in the case of home systems users, depending on their wealth). The supply of local construction materials, labour and transportation are examples of in-kind contributions which the communities made towards the projects.

In both types of projects, the community had to develop the financial means of paying for the operation, repair and maintenance of the systems, and for providing for the possible expansion of system capacity. Revolving funds were the most usual format, and were used to repay the initial loans (Huacaz, 1992, 1994). To date, over 35,000 systems have been installed in Mexico, with some 20,000 systems installed as a result of the PRONASOL programme.

Utilities

In most developing countries, the government is the main provider of electricity utility services. These utilities tend to be under-financed, have inadequate fuel cost-recovery strategies and perform poorly within their sector. Often the electricity utilities do not have enough capacity to supply connected customers on a continuous basis, so their priority of necessity is to increase supply and service reliability to these customers. Those not connected must wait. This typically advances the urban areas, whilst the rural populations remain unconnected to grid electricity.

In Indonesia, the utility is one of the partners involved in the installation and financing of SHS

Photo: Siemens Solar

Even if grid electrification reaches a village, bureaucratic processes and high costs of individual connection often make it prohibitively expensive for local people. For instance, in Sri Lanka, there are approximately 16,500 villages still not connected to the electricity grid. Even in electrified villages, only 10 - 15% of the houses are connected, due to high connection and user costs. This means that around 75% of households in the country have no access to electricity.

To maximise perceived impact and fulfil their own profit and project targets, energy utilities world-wide have traditionally preferred large-scale centralised power systems, such as hydro, diesel generators, gas and coal-fired power stations, to supply power to the grid. Funds for new projects in developing countries have typically come from international banks and funding institutions, and projects have precluded PV on the basis of economics and supply capabilities within the traditional

parameters. Whilst these large-scale fossil fuel projects have in general fulfilled short term objectives and strategies, they have usually left rural communities still unconnected to the national grid, or worse, caused environmental deforestation and relocation of local populations.

There are now pressures being placed upon these utilities to upgrade their performance in order to attract international funding. A number of countries, such as Indonesia and India, have issued invitations to the private sector to participate in the public energy sector. In Africa, many countries are breaking the traditional national utility monopoly and becoming more open to private energy generation. This opens the way for the establishment of smaller, utility-type organisations (Energy Service Organisations), which specialise in the supply and operation of PV and other renewable energy systems (Cabraal et al., 1996).

There are a number of utilities in developed countries (particularly in the USA, The Netherlands, Italy and Japan) who have incorporated PV and other renewables into their energy strategies (Bruno, 1996). Very few utilities in developing countries incorporate RETs into their energy mix. Most of the existing projects were initiated in the early 1990s, as a result of increased awareness of renewable energy technologies, stimulated in part by the Earth Summit in Rio.

An initiative of the Zimbabwe GEF PV project is to develop a programme to install systems through ZESA, the Zimbabwe electricity utility. ZESA will develop a creative approach to hardware and maintenance and build a good fit between solar home systems and its billing and maintenance operations. It will decide whether to lease or sell systems, and has the option to amortise costs over ten or twenty years. ZESA is now looking to see if PV can be integrated into its energy plan (GEF, 1992b).

Stand-alone PV and other RETs systems do not have to be competition for grid extension or a competitive threat to utilities. There are utilities who promote PV systems as part of their energy supply and service programme (such as in Indonesia), as they recognise that

New Approaches to Organisation and Management of Rural Power Supply

In the early 1990s, the Stockholm Environment Institute supported TANESCO, the Tanzanian utility, in an evaluation of the rural electrification programme. The evaluation indicated that: (i) 50% of the rural district centres and less than 1% of the villages had been electrified by 1992; (ii) financial and technical performance of the rural projects was poor; (iii) rural power projects were a big administrative burden to the utility; and (iv) many villages would have to wait indefinitely to be electrified due to limited resources.

In view of the above, a pilot project was proposed which resulted in the formation of the Urambo Electric Consumers Co-operative Limited (UECCO). Urambo became the first village in Tanzania to create an electrification co-operative. The daily management of the co-operative is the responsibility of an elected Development Committee, consisting of 10 people who rotate responsibilities each week. TANESCO provides technical support to the co-operative. The project is financed by IDA. The consumers in Urambo are now receiving a reliable and high quality power supply for 4 hours a day.

A demand-side management programme was also introduced in order to avoid rapid increases in supply needs. Compact fluorescent lights (CFL) were donated by Philips from Holland. Meters were installed later to facilitate the correct billing of the consumers since they used about three times the amount of electricity paid for when the flat system was being applied. Monitoring of the performance of the co-operative is still ongoing. Rural power co-operatives are definitely in line with policies that foster self reliance, and this could be one effective institutional way of organising electricity supply in remote areas. Instead of waiting for help, the rural communities can take the initiative to plan, finance partially or wholly and manage local power projects (Gullberg, 1997).

villages far from the grid will not be electrified in the short to medium term. By a variety of mechanisms, utilities can also assist in the development of the PV market. Utilities can:

- own the RETs systems and rent them to villagers, using the systems as collateral, or part thereof

- use a lease purchase scheme, whereby the householders own the systems at the end of a prearranged period

- provide a mini-grid based upon PV, micro-hydro or wind, operating the system and the service to customers on a similar basis to large-scale electricity distribution

Considering the above three activities, there are also other options for the provision of energy services that include renewable energy technologies, available to utilities operating in developing countries. The suitability of the technologies and options will obviously vary from country to country. Influencing factors will include geographic and climatic conditions, the needs of the population, and the commitment of the utility (and the government) to use non-polluting energy technologies.

Academic institutions

There are a number of important roles which academic institutions can play in assisting the market development of renewable energy systems. In some countries, there are institutions already undertaking activities, such as the Centre de Développement des Energies Renouvelables (CDER) in Morocco. CDER manages the government's PV programme (including international programmes), and provides technical expertise particular in the area of PV. This complements the development of Centres of Excellence. There are other institutions in developing countries undertaking similar roles at a national and regional level. However, to date, most of these activities have been related to technical, not financial, activities.

Activities which academic institutions can undertake which will support the develop-

41

Community street lighting, India Photo: BP Solar

Non-Government Organisations

NGOs can be very important in disseminating information and promoting technology acquisition to grassroots communities. For instance, NGOs such as WaterAid are introducing appropriate, hygienic, sanitary methods for rural villages and training local people to build systems. At the same time, they are educating the people about critical health and sanitation issues. There is a small number of NGOs who are providing financing, or facilitating the flow of capital from international markets to local borrowers.

ment of the renewable energy market within their country and region include:

- training of installers, technicians and users

- training of financial institution staff

- quality assurance programmes for equipment

- Research and Development (as applicable)

- co-ordination of international and regional programmes (in collaboration with government departments)

- information dissemination campaigns

5.3 The private sector

A market-driven approach is the best option for developing any market, including those for renewable energy systems in developing countries. However, for the reasons stated in previous sections, the market for these is not yet developed enough to support substantial industries. Thus other market enablement actions are required in conjunction with private initiatives. There are many actions which the private sector can undertake to develop a sustainable market. Many of these are already being undertaken on a small-scale in developing countries.

The largest and most well known PV NGOs - The Solar Electric Light Fund (SELF) and Enersol Associates - are both based in the USA but operate extensively in developing countries.

SELF has PV revolving fund projects in many countries (Freeling, 1995). For example, in Sri Lanka, SELF and its local partner Solanka have assisted the people of Moropatawa village to form the first 'Solar Co-op'. In Vietnam, SELF has undertaken a pilot project with the Vietnam Women's Union (which represents 11 million members). In Nepal, SELF and its partner, the Centre for Renewable Energy in Kathmandu, have completed the country's first pilot SHS project near Annapurna. This has led to the PV electrification of other villages.

Enersol Associates devised the Solar-Based Rural Electrification Concept (SO-BASEC) which is operating in the Dominican Republic (Smith, 1996). The SO-BASEC model is also being used to provide appropriate financing for SHS in Honduras. Enersol also acts as guarantor for many local revolving funds. For instance, Servicios Sociales de la Iglesia Dominica (SSID) provides PV financing for rural home-owners. SSID manages a $20,000 loan fund, which is guaranteed by Enersol's Fondo Solar.

Enersol rarely directly capitalises a revolving fund, but rather assists local organisations to link up with funding sources. In other instances, it provides loan guarantee support. Enersol established the Fondo Solar Loan Guarantee Fund in 1994, with grant equity from the Rockerfeller Foundation and debt financing from the International Fund for Renewable Energy and Energy Efficiency. The purpose of this fund is to help Latin American organisations to obtain local debt financing for PV programmes. Once the project has been agreed between Enersol and the local NGO, the NGO goes to its bank to negotiate a loan, which Enersol then guarantees. Because the projects are usually in start-up stage (usually around $20-30,000), Enersol usually offers the NGO an institutional support grant equivalent to 10% of the amount of the loan.

A representative from a solar revolving fund talks with a rural home owner in the Dominican Republic   Photo: Enersol

The French NGO, Fondation Energies pour le Monde (FONDEM), has also been the catalyst for the installation of many PV systems in rural areas of developing countries. These include pumping systems, health clinics, schools and battery charging stations, and countries where they have been installed include Vietnam, Senegal, Vanuatu, China, Laos, Cambodia, Madagascar and Morocco.

There are a few development NGOs and Aid Agencies who have incorporated PV and other renewable energy technologies into their programmes. NGOs such as the Aga Khan Foundation, Save the Children, Médicins Sans Frontières (MSF) and the British Red Cross have procured and installed PV refrigeration (conservation of vaccines and blood) and lighting systems for health projects in such countries as Pakistan,

Laos and Zambia. Some also use PV to power communications equipment; and solar cookers have been used in refugee camps to help reduce the need for fuelwood (IT Power, 1996c).

Not many NGOs have made the transition from technical assistance and development to providing successful financial services. Of those who have done so to date, success has been limited. This is due to a number of reasons. Providing financial services means a new operational area with which to become

MSF health clinic, Zaire   Photo: Jean-Paul Louineau

43

familiar and competent. It also often means a change of culture within the NGO itself, going from a 'charity' aspect to the provision of commercial funds. As this Guide discusses, financing programmes should be self-supporting and able to be replicated in another area, in order to bring important financial services to rural areas that are now (usually) devoid of them. Due to their generally strong local knowledge, NGOs are often well placed either to set up appropriate financing schemes or to assist others to do so.

Locally-trained technicians can service local systems and, by doing so, increase the skills base and economic wealth of local communities. Photo: Dulas

There are a number of southern NGOs who specialise in providing financial services to micro-enterprises. These often rely on a system of donor consortia, which makes it easier for NGOs to deal with several donors at once (Wright, 1996). Whilst this system has many advantages, there are also a number of disadvantages, one of the main ones being that funding to the NGOs is usually in the form of grants, and this brings about many of the disadvantages which are discussed within this Guide. However, the concept of 'financial bundling' for NGOs lending on a sustainable basis for RETs purchases, is well worth considering for future developments.

In other instances, a number of NGOs come together, and perform different enabling functions in different areas of the country. For instance, Friends of Women's Banking-India (FWWB/India) is an umbrella for a network of NGOs working to promote development through access to credit and savings for the rural poor. It has directly assisted over one million families. FWWB/India, through its member NGOs, promotes the establishment of informal, village-based savings groups. These savings groups are given training and support to become self-reliant and self-managed, with regular savings and an efficient structure. The financial discipline and regular savings of these informal groups have enabled them to take out loans with banks (WWB, 1993).

Private companies

The role which private companies and businesses can play in the implementation of renewable energy programmes is very diverse. In general, the role of companies and businesses has been to manufacture and supply equipment, not to provide loans to distributors or purchasers. Table 5.1 gives an overview of the structure of the PV industry. This structure affects the private sector's ability to supply equipment, particularly SHSs to point of use in rural areas, and to sustain a maintenance and spare parts infrastructure in these areas. At present the world PV market is dominated by a few large companies who have manufacturing, supply and (though not always) installation divisions. There are increasingly smaller distribution and installation businesses setting up locally. If this trend were to continue, aided by international technology transfer programmes, then there would be more expertise at a local level. This pattern could be reflected successfully for most renewable energy technologies, with some businesses providing a range of renewable energy technology services and others specialising in one or two technologies (depending on local demand).

At present, world-wide the private sector has only a fragment of the services it will ultimately provide. The reasons for this vary, and are affected by country and regional dichotomies. However, one of the main contributing factors for a small private sector in renewable energy technologies is the small size of the market. Once a critical market mass has been established within a country, then a plethora of private initiatives will develop. There are a number of national initiatives, such as those funded by the GEF, which are trying to overcome this problem by assisting market development.

Services that the private sector can provide relate both to the technology and to financing the procurement of the technology at a local level.

At a technology level, there are hundreds of companies and businesses world-wide which make and/or supply renewable energy systems. However, in developing countries, rural infrastructures for supply of equipment and spare parts are poor. This is due to the small existing and generally less accessible market in rural areas, and to the lack of appropriate financing available to businesses to allow for their own expansion and investment. Table 5.1 profiles the diversity of the existing PV industry.

Technical services that can be (and, in many cases, are already being) provided include:

- manufacture, supply and installation of equipment

- maintenance and after-sales services

- training of installers and users

At a financial level, there are not many organisations providing finance for renewable energy systems at a local level (refer to chapter 6). This needs to change, and as it does, the market will grow and will become more profitable for financial organisations. Two important financial services that need to be provided are:

- dispersement from large international and national loan funds to small, dispersed local funding schemes. This may be undertaken by an intermediary organisation (e.g. Solar Electric Light Fund) or directly from a national organisation to local lending (e.g. IREDA)

Table 5.1 Profile of the PV industry

Organisation	Activities	Typical size	Advantages
International and regional companies	- manufacture cells - manufacture modules - manufacture BOS components - distributors of systems - distributors and installers of systems	- 50 - 500 employees world-wide	- large-scale manufacturing - wide network for equipment distribution - in-house expertise
Local companies/ organisations	- system distributors - distributors and installers of systems - installers - local manufacture of BOS components (e.g. batteries, lighting kits) - module assembly - cell	- 1 - 100 people	- local market knowledge - BOS components can usually be manufactured more cheaply at similar quality

45

Country and Programme	Government	Utility	NGO	Bank	International Agency	Estimated number installed	Technical component	Financing scheme used
Mexico: PRONASOL and private sector initiatives	✓	✓	–	✓	–	35,000+	✓	Government programme, involving local credit co-operatives
Indonesia: BAN-PRES and other pro-grammes	✓	✓	–	✓	–	40,000+	✓	National bank, providing funds for revolving funds
Dominican Republic: SO-BASEC and ENERSOL projects	–	–	✓	–	✓	5,000+	✓	International funds providing seed money for revolving funds, involvement of NGOs
Zimbabwe: GEF	✓	✓	–	✓	✓	2,000+	✓	International finance for revolving schemes
India: GEF	✓	✓	✓	✓	✓	?	✓	International finance for preferential lending
Bolivia: Altiplano region	–	–	✓	–	✓	1,500	✓	Locally funded credit co-operatives
Tuvalu: TSECS	✓	–	✓	–	✓	1,500+	✓	Energy Service Company, utility and French funding
Sri Lanka: Savodaya	–	–	✓	–	✓	100's	✓	International funding for local revolving funds
China: Gansu SELF and other programmes	✓	–	✓	–	✓	20,000+	✓	International funding for local revolving funds, government programme including private financing
Kenya: Private market development	–	–	–	–	–	20,000+	–	Primarily cash sales
Morocco: PPER, SEP, EDF programmes and other mechanisms	✓	–	–	✓	✓	5,000+	✓	International funding for revolving funds, bank financing and some grants

Table 5.2 Summary of national SHS programmes

- management services (i.e. running local revolving loan funds, credit co-operatives, and so on)

Much of the SHS world market (in particular) to date has been the result of interaction with government of international programmes. However, there are significant country markets, and a huge unfilled potential (Cabraal et al, 1996). As stated previously, lack of financial mechanisms is one of the most cited reasons for slow development.

The private sector can play a significant role in assisting the development of both funds and the infrastructure to support them.

i.   *Financial*. International and national organisations can provide funding for the following activities:

- development capital
- fund capital
- training bank staff
- financial services (e.g. hire purchase)

ii.   *Technical*. There are many roles which international, national and local organisations can play to ensure that quality products are available on the market:

- supply of quality equipment
- installation, repair and maintenance services
- local manufacturing and equipment assembly

Technical services can be either integral to the fund and its operation, contracted out from the fund, or work independent to the fund. The role of the private sector's interaction with the fund needs to be defined before the start-up phase, so that their requirements for profitability can also be taken into account. Two important issues for the sustainability of the programme and its support structure are adequate supply of equipment and density of installed systems.

5.4 Conclusions

When investigating possibilities for the establishment of financing renewable energy technologies in a developing country, it is vital to look at the international, national, regional and local institutions, and ascertain the impact which they could have on the proposed scheme. Table 5.2 summarises the main PV programmes which have happened in developing countries in the last five years.

In many cases, the activities which different organisations can undertake overlap. This is not a problem to the development of the market within the country, as the more organisations that are active within the sector, the larger the sector should become to support them. Overlapping only becomes a problem when different views are expressed by different organisations, and this acts to disrupt market confidence. If government infrastructures are in place, then any differences should be able to be arbitrated easily by a non-biased institute of expertise.

This chapter has looked at the role which institutions can play in setting up financing schemes for renewable energy technologies. In the next chapter, different types of financing mechanisms are discussed, and in chapter 7 potential funding sources are explored.

Financing mechanisms 6

As chapter 3 demonstrates, renewable energy technologies in general suffer from high initial costs compared to other energy technologies. While life-cycle costs may be far less than those of conventional energy options (which generally incur higher operation and maintenance costs), RETs' high capital costs have tended to price them out of the range of many potential users, particularly low-income rural households.

To make the majority of the renewable energy technologies more affordable for more people generally requires costs to be spread out over a greater time-frame. This is where credit brokers and long-term repayment schemes can play an important role.

This chapter considers the various actors - be they individuals or organisations - involved in the provision of financial assistance for rural people in developing countries. The roles which these actors can play in developing small-scale sustainable financing is discussed in more detail in chapter 5. Some mechanisms which have been, or could be, adapted to finance the sale either of renewable energy technologies themselves, or the energy service made available through the use of RETs, are also presented in this chapter.

6.1 Lending schemes

Indigenous lending schemes

All indigenous cultures have some type of local financing which has come into being as a result of local demands. In developed countries this operates through a network of commercial banks, building societies and other formal structures. In many developing countries (outside the main urban areas), less formal financing systems tend to predominate. This is mainly because people living in rural areas tend to have less capital and access to wealth creation, and therefore there is less requirement for large wide-scale lending.

All costs must be allowed for including transportation to remote areas

Photo: Intersolar

48

Typically, borrowers in these centres only want to borrow small amounts at one time, which is not cost-effective for the banking institutions with large overheads to sustain.

Indigenous lending systems vary from country to country, and region to region. Their common denominator is that they have evolved to suit the specific needs of the people in their lending region. Lending variables have to take into account the seasonal nature of incomes derived from farming and livestock breeding, droughts and other natural disasters, economic fluctuations both regionally and nationally, wealth creation possibilities of the local people and so on.

The following are examples of such existing lending organisations. Some have already proven their suitability for financing the purchase of small renewable energy technologies, whilst others could well be adapted for this purpose.

The money lender

At one extreme is the traditional money lender, who charges very high interest levels even for short-term loans. He/she is often maligned but has made a positive (though not quantified) contribution to development through supplying much needed credit in areas not served by formal institutions.

Money lenders and other informal credit structures exist because they perform a function not otherwise fulfilled in a society, and come in different forms, depending on the society and the transactions required. It is possible that money lenders could provide financing for RET purchases, but it is not likely that this will be formalised. Rather, success with lending through the money lender in a given area, interest rates charged, payback rates and so on can be used as an indicator of borrowing experience and reliability in that area.

Traditional credit organisations

Traditional credit organisations in Africa include *susus* in Ghana, *fingongs* in northern Tanzania, *equbs* in Eritrea and *tontines* in Senegal and Rwanda. Bolivia has the *pasanaku*, the Dominican Republic the *socio*

and Haiti the *solde*. These are all informal savings organisations which can include both rotating and accumulated savings and credit. They allow members to save small amounts of money and can have pre-arranged assignments for loan amounts.

Regardless of their structure, they have legitimate and established contracts into which people enter for their financial needs, and they fit into the social and cultural norms of the societies in which they operate.

Their main advantages are that they are adaptable to local situations, they are not costly to run (small, if any, overheads), can rotate lending daily, do not require loan recovery and reprocessing procedures, and often combine savings with credit. However, one disadvantage is that the savings do not earn any interest.

Banking

Commercial banks

In general, privately owned commercial banks have not been interested in financing small decentralised renewable energy systems. Such loans have high transaction and administration costs and are largely considered too cumbersome and low profit for traditional banks to pursue.

Commercial banks also tend to undertake detailed client screening procedures and will require substantial collateral as security against failure to repay the loan. Land is commonly used as collateral, but many people in rural areas of developing countries may not own land, or may not be able to provide proof of land ownership.

In Europe, for example, banks often require material security of 100-120% against a loan. In Latin America up to 300% collateral might be required. So even if the customer would have no problem in meeting the repayments, they might be unable to take out a loan for want of acceptable collateral.

Low literacy and numeracy skills, which can result from a lack of formal education, also pose a problem. Bank loan application forms are designed to safeguard the bank as much as possible from repayment defaults, and can appear complicated and confusing. Typically, there is also a lack of familiarity with and/or ability to carry out the financial obligations which the banks require. These are significant deterrents for many would-be applicants.

Often then those most in need of assistance, having insufficient credit history and little or no collateral, are viewed as too high risk. They would therefore normally be ineligible for a loan from a traditional bank.

Occasionally ways of circumventing traditional collateral requirements are found. For instance, savings that are returned to the customer once the loan has been fully repaid may be an acceptable form of security. Alternatively, the reason for taking the loan, i.e. the equipment that is purchased, could be accepted as collateral (in much the same way that a house is security against a mortgage). However, whereas the equipment would have a resale value for a vendor, this is generally not true for a bank, so additional collateral will normally be needed.

Development banks

Development banks have a different objective than their commercial counterparts. Usually government controlled, they tend to have an underlying political agenda which may or may not bring benefits to the population they serve.

However, like the commercial sector, development banks are not generally in the habit of disbursing small loans (e.g. to individual rural households) unless it is politically appropriate to do so.

Interest charges on loans that are granted will usually be very low - possibly negative in real terms. Furthermore, repayment collections are not always strictly enforced which does little to encourage financial accountability amongst borrowers and is not an incentive to break away from central government 'handouts'.

For poorer households, then, the formal financial sector is invariably inappropriate as a source of direct credit. However, as the accounts alongside illustrate, there are some

> ### People's Bank, Sri Lanka
>
> In Sri Lanka a loan from the People's Bank project finances up to 80% of the PV system cost, with 18% interest on the loan per annum over a maximum lending period of 60 months and a strict requirement for loan collateral. Before being granted a loan, the borrower must prove that he/she has a regular income and a good social standing (IT Power and Eurosolar, 1996).

notable instances of banks successfully lending to rural peoples for life-improvement projects involving RETs, such as PV for lighting houses or community centres by the Sri Lankan People's Bank.

Over and above the cases outlined here, the Grameen Bank, although not specifically designed as a means of financing renewable

> ### L'Union des Banques Populaires du Rwanda
>
> L'Union des Banques Populaires du Rwanda provided credit for the purchase of PV systems in rural Rwanda. Here, where an initial lighting kit costs around $560, the local people need some form of appropriate financing to be able to purchase the equipment. The conditions for the loans were that:
>
> - there were at least five interested persons per village
>
> - each person put up 20% of the system costs
>
> When these conditions were met, the bank paid the total cost of the system to the supplying company, who then proceeded with the installation. The reimbursement of the loans was spread over 12 to 37 months, with instalments ranging from $14 to $40 per month and an interest rate of 9%. As a result, 1450 lighting kits were sold and installed in one year alone (Louineau et al, 1994).

energy projects, is probably the best example of one bank's initiative to provide unsecured loans to landless householders (see 'Peer Group Lending' later in this chapter).

There is also scope for banks to be brought into the equation by intermediaries (e.g. commercial RET equipment dealers, NGOs, solidarity groups etc.) who can co-ordinate refinancing of many small loans from a few large bank loans.

Community bore-hole pumping system, Morocco   Photo: BP Solar

Field officers and intermediaries

Field officers and other intermediaries can act as a link between institutions and rural clients by performing a range of functions such as identification of borrowers, preparation of bankable schemes, disbursement of credit, assistance in the productive utilisation of credit, collection of repayments etc. Field officers can be attached to a specific institution, act as independent intermediaries or represent a number of financial institutions at the same time.

Intermediaries are either organisations or individuals who commonly retail credit from larger commercial banks to small borrowers. They finance their operations by charging the borrowers a small fee for their assistance in facilitating the loan. One such intermediary is Women's World Banking, which secures small loans for women's projects in developing countries from larger formal institutions and oversees the repayment of these loans.

Field officers and other intermediaries can assist in reducing transaction costs to lenders as they operate on a community level, tend to know the potential borrower and/or have ready access to information about the creditworthiness of such, and have low overhead costs.

The most effective field officers have usually received hands-on training relevant to actual field experiences as well as lending procedures. Such field officers have been operating in rural India and Africa for many years, their relative success being directly related to the quality of training and motivation which they receive. Apart from a

Dominican children pose under their PV panel, which provides lighting and TV in their home. This was financed through La Asociacion para el Desarrollo de Energia Solar   Photo: Enersol

basic knowledge of banking, accounting and financial management, many are also trained in elementary technical knowledge.

This is the case with the mobile credit officers in the Agriculture Development Bank of Pakistan (IT Power, 1996c), who operate from branch offices and cover about 25 villages each. In Africa, Fonds National de Développement Rural has utilised the field staff of government agencies for distribution of inputs and recovery of credit. A small fee is paid to these agencies. This arrangement is more economical for the bank than training and sending out its own field officers.

Another variation is the 'one man village banker scheme' in Pakistan (IT Power, 1996c). A resident of the village is appointed by the financial institutions, and he is accessible to clients at his residence throughout the day. This is closer to the concept of the local money lender, and may be less daunting for first time borrowers than more formal institutions.

Credit co-operatives

Credit Unions or Credit and Savings Co-operatives are already in existence in many rural areas, traditionally to provide finance for housing, agricultural supplies and equipment. These have a very localised focus, both in the respect that they are based on local requirements and local ability to repay, and that they mobilise purely locally generated resources, rather than channel finance from outside. Credit co-operatives are often small in size with a typical membership of no more than fifty. Each member within a co-operative will be known to at least one other member. Peer pressure therefore plays an important role in assuring loan repayments.

Before gaining eligibility for a loan, however, new members must first buy shares in the co-operative by depositing savings, and must continue to demonstrate their ability to maintain a savings programme. This serves as a basic screening process, but may also be a useful reserve to guarantee, at least in part, the eventual loan repayment.

After several months members may take out loans, if approved, of up to three times the

TSECS, Tuvalu

The Pacific country of Tuvalu consists of three islands and six atolls. It has a population of about 9,000 (1991) almost half of whom live on Funafuti, the country's administrative centre. Of the 1,000 outer island households over 400 now use PV for lighting.

Since 1984 solar home systems have been installed through the work of the Tuvalu Solar Electric Co-operative Society (TSECS). The Society was set up by the Save the Children Fund as a commercial enterprise to implement and manage a rural lighting project. It received its first 170 systems with USAID funds which have since been supplemented with grants from the European Community and French Government to add battery controllers, replace batteries and provide an additional module to some systems which were initially found to be under-powered. Despite many technical problems during its first ten years, the TSECS has continued to provide maintenance and spare parts for its 400 systems.

The systems are owned by the TSECS and participating households have a shareholding in the Society. Co-operative branches have been established on each island after at least twenty households paid a deposit of US$36.40 and were prepared to pay a monthly service charge of US$4.55 to cover maintenance, organisational costs and spare parts. Although households do not fully repay the costs of their systems, the TSECS could provide a useful model of a rural electrification project utilising a revolving fund and actively involving the community.

The success of the TSECS is thought to lie in its organisation which is both community based and has access to external resources. Availability of technical assistance is also seen as a major factor.

A project based on the TSECS model has recently been set up on the neighbouring Pacific island of Kiribati and is performing well (repayment rates are over 95%) (Conway and Wade, 1994).

value of their savings. Members can only claim their savings by leaving the union.

In 1989, there were 17,000 credit unions in developing countries (Otero & Rhyne, 1994) with 8.7 million members. Because their operational costs are low, provided repayments are received on time, interest charges alone can be enough to allow credit co-operatives to sustain their activities without external contributions.

The interest paid on savings is in the form of dividends which are usually low, and negative in real terms. Through this mechanism, savings are not a commercially attractive option in themselves but provide a means of obtaining credit. Because credit co-operatives are heavily reliant upon their members' savings as the major source of funds, the lack of incentive to save can limit the available credit resources. This in turn can tend to restrict further growth. More favourable returns on savings could help to overcome the problem. This would only be possible by increasing borrowers' interest charges.

There are a number of credit co-operatives in developing countries which, through their interaction with organisations from developed countries, have been able to adapt their lending to cover small PV systems.

The US National Rural Electrification Co-operative Association (NRECA) is assisting communities in the Valles Crucenos region of Bolivia to access PV lighting systems, as part of the Electrification for Alternative Development project (Aguilera and Lorenzo, 1995). Initially, NRECA has donated 200 PV lighting systems to a number of coffee and electricity co-operatives, who have installed them for customers. The revenue from sales of these systems will be used to finance the acquisition and installation of other PV systems.

In Sri Lanka, the Solar Electric Light Fund (SELF) and its local partner Solanka have assisted the people of Moropatawa village to form the first 'Solar Co-op' in Sri Lanka (Hankins, 1993). Funded by SELF and the Moriah Fund, this now funds over 48 PV home systems. Solar Co-op members pay a small down-payment and low monthly installments at reduced rates into a locally-managed revolving fund. As the fund grows, the number of financed installations will also expand. The Co-operative employs a local technician trained by SELF and Solanka.

Non-Government Organisations (NGOs)

Prior to 1990, the ethos of development NGOs as humanitarian aid organisations meant that whilst they were invaluable in channelling money - aid grants, for instance - they were not on the whole successful financial managers. Charging interest on loans or otherwise acting in a commercial way was seen then as undesirable by the NGO community. As a result, operational costs could not be covered and many NGO credit schemes proved largely unsuccessful.

NGOs do have a potentially important contribution to make in providing credit for financing purchases of renewable energy systems, but the traditional role as outlined above needs refining.

In southern Democratic Republic of Congo, MSF has undertaken rehabilitation of PV systems and training of health staff how to use their PV systems. This includes instruction sheets mounted beside the lighting kit. This was financed through MSF for local people   Photo: Jean Paul Louineau

53

NGOs are now recognising that by improving the quality and efficiency of their financial services they can mobilise substantial domestic savings and so assist many new borrowers. These finance-oriented organisations often offer both loan and savings facilities coupled to realistic rates of interest. Savers therefore have greater incentive to increase their savings whilst borrowers recognise that the agreement they have entered into is serious, which helps to limit any deliberate intent to default. With this new role of financial intermediary which has efficiency at its core, the finance NGO is perhaps more comparable to a credit co-operative than it is to the commonly held view of an NGO.

Community Street lighting, Brazil
Photo: Newcastle Photovoltaic Applications Centre

Obviously to achieve such a change in function requires considerable structural and operational reorganisation. To this end, recruitment of the appropriate specialist staff and some retraining of existing staff will be necessary. Major investment in suitable finance information and management systems is also unavoidable. These additional costs will, by necessity, be transferred to the end users. It then becomes increasingly important that efficiency is maximised in other areas such as applicants' review and approval procedures, loan disbursements and repayment collection arrangements .

6.2 Making RETs affordable

It is difficult to specify a price boundary below which RETs become 'affordable'. This depends very much on the disposable income of each customer and on a somewhat subjective assessment of what the goods (or service) being purchased will add to their quality of life. In addition, affordability does not necessarily remain constant, but can vary from season to season, month to month or even daily.

Rural farmers offer a prime example of a sector of society which experiences significant variations in wealth over the period of a year.

Assuming crops are successful, they will typically have access to a comparatively substantial cash resource shortly after harvest, but may have little at other times of the year. In such cases, it is quite possible that a renewable energy system would be purchased for cash when the resource is available, unless a credit arrangement which accounted for such monetary fluctuations were available.

Other potential users, who have access to adequate savings and who feel that a particular renewable energy system will improve their lifestyle and/or increase their earning potential, might also consider a straightforward cash sale to be affordable. Indeed the majority of sales of Solar Home Systems in both Kenya and the Dominican Republic have, perhaps surprisingly, been paid for in a single lump sum.

Often though, savings will be insufficient to enable an outright cash purchase. So for many potential customers, longer term financing schemes which offer a mutually acceptable repayment schedule can improve affordability. If renewable energy services are to be made more widely available to rural populations such alternative credit lines are essential.

From the customers' perspective, the main alternatives for longer-term financing of RETs fall into four categories:

- consumer finance (short to medium-term loans)
- lease financing
- service contracts
- rental schemes

Consumer Finance

Hire Purchase

Like retailers of any other products, dealers and distributors of renewable energy systems are, by definition, intent on securing sales of their merchandise. The hire-purchase (HP) arrangement is the vendor's method for making expensive goods affordable to customers who cannot manage a single cash payment.

Vendors who have good credit-standing can act as a credit broker between their customer and a third party source of finance, typically a commercial bank. Effectively, the bank pays the vendor up-front on behalf of the customer. Ownership of the system is immediately transferred to the customer who makes regular repayments to the bank through the dealer. The dealer will charge a commission to the customer for facilitating the loan.

Hire-purchase dealers tend to be restricted to a small client base. For the customer this has both advantages and disadvantages.

On the negative side, commercial dealers are not well placed to take advantage of bulk purchase discounts and are unlikely to be able to arrange favourable financial terms from banks or other sources. As a result, this will be reflected in relatively high charges to the customer, both in terms of the initial deposit and the interest charges.

However, commercial vendors have a strong incentive to ensure a sale and can avoid much of the bureaucracy of larger organisational structures. They can therefore provide a far less cumbersome loan processing procedure.

As ultimate guarantor of the loan, it is not in the vendor's interest to sell to high risk customers. The equipment will normally be held as collateral, and the customer will generally be required to make a large downpayment to safeguard the vendor. The reasoning for this is that customers are less likely to default on payments if they have already made a significant financial investment in the equipment.

Peer Group Lending

Another course for obtaining a loan is through a peer (or solidarity) group. This assumes that a suitable group lending programme is operating in the target region. Typically this would be managed by an NGO or a financial institution specifically geared to poverty alleviation.

The arrangement relies on a kind of natural selection procedure whereby borrowers form their own peer groups - typically consisting of between three and ten friends, relatives or colleagues. The group as a whole can then approach an appointed field representative from the programme management organisation who will determine whether or not the loan should be granted.

The arrangement is attractive in that solidarity groups are largely self-screening. Because the group as a whole is responsible for ensuring prompt repayments it is in the individual member's best interest to ensure that their partners are suitably responsible. The members assess whether or not an individual's request for a loan should be taken up by the group. As the group and the individual members will only be eligible for additional loans if the repayment schedule of the current loan is adhered to, peer pressure helps to minimise the lending risk. The associated benefit of this arrangement for the programme managers is that the costs per borrower of administering the loan are reduced.

Nevertheless, it has been estimated that the cost of operating a solidarity group programme can amount to 25 to 50% of the total portfolio (Otero & Rhyne, 1994). To guarantee the sustainability of the programme these costs must be reclaimed through borrower service charges and interest rates. There may also be a degree of compulsory saving which serves two useful functions: from the programme managers' perspective these may contribute to the programme lending funds, which reduces the need for external contributions; from the users' perspective they can serve as a fall-back for instance if a repayment goes awry.

The peer group mechanism can have additional social benefits as it revolves around a concept of mutual respect and trust - both between individual group members and between the group as a whole and the programme administration. For the participants, the peer group experience is a significant move towards some financial autonomy.

The Grameen Bank, which was established by the Central Bank of Bangladesh in 1979 as a project in experimental credit to provide unsecured loans to landless householders, is one of the best reported accounts of a successful Peer Group Lending programme.

Leasing

For the user at least, the leasing and service arrangement can be quite similar to each other. The customer signs a contract with an agent - possibly a utility, co-operative or NGO - who brokers low-cost credit on their behalf from a bank or similar source of finance, and arranges installation (and possibly maintenance) of the RET system.

The lessor is responsible for screening customers to determine their ability to meet the proposed repayment schedule, and guarantees loan repayments to the bank.

Because lease agents manage the finances for a large number of clients, they may gain considerable bulk purchase discounts from equipment suppliers and favourable terms for agglomerated finance from the bank.

The consumer makes regular payments to the lease agent for hire of a system and for its installation. The payments may also cover service of the system, depending on the conditions of the agreement. If service is not specifically included in the contract it will be left to the customer to arrange for the system to be serviced (and repaired as and when necessary), usually by an independent contractor.

The lessor will own the system throughout the period of the agreement, but eventually, the ownership title may be transferred to the customer. This is the fundamental difference between the lease arrangement and the service contract (see below) where the user never actually owns the system.

If the lessee does not keep up payments, the financing institution has the option to repossess the module and other removable components. As the PV module has a lifetime of upwards of 20 years it can be used as the collateral for the lease.

Service contracts

The service mechanism is closely related to the lease model except that the end user never actually owns the system but essentially hires it from a service organisation over an extended period.

This eases fears surrounding the credit-worthiness of the customer as the service organisation always has the option to repossess any 'removable' equipment of value (e.g. the module and battery from a PV SHS) in the event of default on payments for subsequent reletting. Deposits for systems hired under service contracts will generally be lower than commercial loans, but should be sufficient to cover the cost of any non-recoverable components.

Repayments can be set so that costs can be recovered over the expected lifetime of the equipment. This could be as much as ten or even twenty years, which will translate to lower monthly payments for customers (when compared to purchasing the system with a commercial loan, typically to be repaid over two or three years). However, it is important to remember that there are additional recurring costs in this model (i.e. continual maintenance and probably also some component replacement) which can be avoided with other arrangements.

The basic concept which the service contract model revolves around is, as the name suggests, in the equipment servicing arrangements. The onus is entirely on the service organisation to undertake maintenance of the system on a regular basis.

This is an additional cost which will be transferred to the customer through the payment schedule, but it has distinct advantages both for the provider and the recipient. For the

Renting schemes can be used for a variety of renewable energy products, such as solar cookers   *Photo: SEHUF/Synopsis/VTZ*

The down side is that if such projects do not at least break even, the draining effect of providing replacement components and service under the agreement has the potential to spiral out of control and quickly bankrupt the fund.

6.3 Revolving funds

The previous section outlined several mechanisms that could make RETs more affordable. This section attempts to illustrate how these models can realistically be implemented.

customer it means that they can receive all the benefits of the energy system safe in the knowledge that, even in the event of equipment failure, they will not be without power for long periods and they will never have to pay vast sums for repairs needed as a result of fair 'wear and tear'. The service organisation, on the other hand, can be confident that maintenance is actually being carried out, so if repossession should ever be necessary their equipment will be functioning correctly.

Renting schemes

Renting a PV system can be an attractive option in that it offsets all capital costs and avoids financial contractual obligations. However, there can be little sense of ownership with a rented system, and maintenance and any repairs may not be undertaken, thus reducing the life of the system. Yet renting schemes for PV systems have been successfully implemented as the Senegal example demonstrates.

If the correct balance is struck, the service and rental arrangements can provide a prime opportunity for exponential fund growth. This is possible because every participant in the scheme (barring those that default on payments) is providing continuous fund revenue. If a profit is made on each system further customers can be financed.

Consumer financing accomplished through commercial dealerships is largely beyond the influence of the development worker. However, there is still an important role in raising awareness of credit availability and finance terms, and in providing the educational and informational support to ensure that the potential customers recognise the social and long-term economic benefits of RETs.

Outside the commercial sector, though, there are considerable opportunities for long-term renewable energy development project initia-

> **Village battery recharging stations, Senegal**
>
> In Senegal, 36 systems, comprising a PV recharging station with five PV portable lamps per system, have been rented to small villages since 1991. The project is managed at the central level by the Centre d'Etudes et de Recherche sur les Énergies Renouvelables de Dakar and at the local level by existing local organisations. Between 10 and 15 lamps are rented by villages at a cost of $0.38 for each recharge. Despite some technical problems due to the poor design of the lamps, rental demand generally averages around 80% of the supply capacity (Louineau et al, 1994).

tives. The revolving fund can be an important mechanism for implementing such measures.

What is a revolving fund?

In the broadest sense, all of the credit mechanisms detailed above adhere in some way to the revolving fund concept. The basis of the revolving fund is that an organisation (or an individual in the case of the money lender) has access to a reserve of money (funds) which is used to lend to one or more borrowers. Over a given period of time the borrower is expected to repay the original sum which replenishes the fund. This money can then be loaned to a new borrower and so the cycle starts afresh. Usually an additional sum (interest) is charged to the borrower by the lender, which effectively serves as a fee for providing the service and helps safeguard the fund from factors which could deplete it (e.g. inflation, non-payments and the cost for the lender to obtain external finance). Revolving funds can serve to mobilise domestic savings by encouraging users to invest in goods or services which they pay for over an extended period.

The credit co-operative for instance is a non-formalised revolving fund. Members deposit savings which are held in a common fund. These are lent on to other members when the need and availability arises. The loans must be repaid so that the original fund is refilled and the lending process can begin again.

Really the difference between such informal rotating credit schemes and the commonly held view of the formalised revolving fund lies in the source of the initial fund, the scale of the operations and the structure of the credit management.

Typically, formal revolving funds are started with seed money from an outside agency (or agencies) and are organised by an existing or new local organisation. The seed is used both to set up the project operational structure and to loan out to many small borrowers. Repayments by the original borrowers replenish the fund enabling further lending to other borrowers. If correctly implemented, revolving funds can be an excellent means of making affordable credit available to the rural poor with a relatively small seed fund servicing a relatively large population.

Seed funding is usually a one-off subsidy to initiate the revolving mechanism. This is preferable to continuous subsidies which have shown a tendency to distort market prices and have in the past proven difficult to remove once customers have become accustomed to them.

Historical perspective

Formal revolving funds are the creation of development programmes. During the 1960s permanent revolving funds were tried by national and international agencies and governments for use within the agriculture sector. The premise for these was that money was loaned to the central bank of the country, which then lent on to commercial or development banks, who then loaned either directly to rural producers or to a further intermediary. These were primarily loans for agricultural production.

The expectation was that, through the provision of this loan facility, charging lower interest rates than otherwise commercially available, the farmers would be able to increase their profitability and repay the loan. The interest charges were thought enough to pay for all the institutional costs. However, by the 1970s, about US$90 billion was outstanding world-wide in rural credit (Otero and Rhyne, 1994), caused mainly by:

- inability to cover institutional costs - the total income gained through charging interest to borrowers was not sufficient to cover the costs of operating the programme (e.g. paying staff to process the loan and collect payments, covering fund devaluation caused by inflation, maintaining offices and equipment, providing for bad debts etc.)

- deviation of funds - the money was not getting to the small rural farmers for whom it was intended. Instead it was being 'hijacked' by the more wealthy and financially-aware sector of the rural community.

More recently, revolving funds have been established which charge near commercial interest rates, and these are proving more effective.

6.4 Conclusions

It is clear then that there are numerous actors and various mechanisms which could be, or have been, adapted to finance renewable energy projects in developing countries. USAID's 1972 review of rural credit institutions (Otero and Rhyne, 1994) highlights the key points to remember from this chapter:

- no particular type of institutional system is applicable for all situations in all countries

- successful institutions are those most able to reach the maximum number of people with the minimum costs

- group activities with an element of compulsory participation have a greater degree of success

- shortage of local leadership and fear of local or central government manipulation acts against many local co-operatives

- all rural credit agencies within a country, regardless of type of organisational structure, should be co-ordinated as a mutually supportive system, retaining their individual attributes but able to share lessons learnt from problems and successes

- rural financial operations must be viable

- trained professionals are essential for lending, understanding local issues (such as seasonal ability to repay loans) and technology requirements

- there needs to be self-evaluating mechanisms to monitor progress and review achievements.

Source of funds 7

As discussed in previous chapters, there are a variety of options for accessing funds in order to start up or supplement a scheme for RETs financing. The source of funds has a critical impact upon whether a fund is managed and how it is able to achieve self-sustainability.

Funding for revolving loans and other financing schemes can come from three sources: bank funds, client savings and third party financing. There is no right or wrong way to mobilise funding and give services to clients, provided that certain criteria are met.

Issues vital for a sustainable fund include:

- Minimal risk exposure. This is important especially for a new fund, as there is no history of repayments or fund management, and thus the risks for lending to the fund are greater. If this risk is spread between a number of financial organisations (for instance, the revolving fund and the bank from which it is lending), then each will incur less risk than if financing alone. This also means that there will be two organisations scrutinising the success of the fund's programme, which usually would ensure sounder financial management.

- Fund lending at the best rates possible. If the risks are perceived as high, the interest rate is usually set high to cover possible defaults. If there is an established pattern of savings per lender participating in the fund or a mandatory savings requirement with the loan, then the risk is minimised and the interest rate does not have to be as

high. Loans to the fund can also be secured from primary lenders (for instance, ethical investment funds) who have lower required rates of return than commercial banks.

- For small funds in particular, banks can provide efficient management expertise and an experienced savings function. Shared activities can include the complete lending function of the fund, or specific activities such as primary loans to fund deposits and disbursements.

- Enough earning on the fund to cover operating costs and risk premiums and to increase the fund size.

There are a number of international financial organisations that will provide loans for RETs lending within a country. These include the World Bank and the International Finance Corporation (IFC). These are dealt with in more detail in chapter 5.

7.1 The financial sector in developing countries

The choice of lending mechanism will vary from country to country, affected by regulations and the development of the financial sector. Developing countries share a number of characteristics in the development of their financial sector. Whilst there is disparity between the regulatory controls imposed by individual governments, the nature of the informal sector in particular has similarities.

There are many thousands of rural lending organisations in developing countries: *tontines*, *djanggi* (Cameroon), *ekubs*, *idirs* (Ethiopia), *tandas* (Mexico), *chit* funds (India), and so on. These have often had an integral role in the development of communities (particularly in rural areas) over hundreds of years. The size of the informal sector within a country can be significant. An OECD study in 1991 estimated that the mean size of the informal sector within a developing country is around 70% of Gross Domestic Product (GDP) (Germidis, 1991).

However, because the majority of these funds are within the informal financial sector, they are usually not regulated by the government (with the notable exception of India). Being outside public control, they are able to overcome many of the bureaucracies imposed on standardised financial transactions, and are more flexible in their lending. On the other-hand, they do not participate in official monetary and credit policies, and can even curtail economic development through high charges on consumption loans (Germidis, 1991).

The formal financial sector, in areas where it does service the rural populations, usually siphons savings into urban areas, where it has

This solar lantern used in Senegal to provide light for cooking is lower in cost than a SHS and could be purchased by either cash or a financing scheme
Photo: Neste Advanced Power Systems

the greatest number of borrowers, rather than using them to develop rural areas. In doing so, it depletes the rural areas of vital local equity for development.

There have been many attempts to create new financing organisations in rural areas of developing countries. However, many of these have not worked. Much can be learnt from the last 20 years of micro-enterprise and agricultural lending.

There are many different types of financing schemes that are applicable to facilitate the purchase of PV systems. Whilst experiences show that there is no one scheme that will be applicable for all situations and countries, there are commonalities.

Formal sector

The formal financial sector is regulated in all countries, though to varying degrees depending on the government. The more the regulation, generally, the less the formal sector is able to reach large numbers of small savers and borrowers. This is largely due to more bureaucracy, higher bank administration costs, and higher levels of literacy and financial understanding required on the part of the customer. Poorer people are often afraid to use banks because of their low literacy levels, which has caused them to be unfairly treated by bank staff (WWB, 1993). Usually, banks have a minimum savings and loans threshold, which is above the limit of poorer potential customers and this also acts as a barrier to small-scale financing. Restricted hours of trading and centralised operations in large urban areas also affect the ability of particularly the rural people to access the services offered by banks.

However, there are organisations within the formal sector for whom RETs lending could be both appropriate and profitable. Building Societies, for example, disperse millions of dollars in loans for housing-related activities. These could be extended to include SHS and other applicable RETs. The Grameen Bank has been used as a model to set up a number of similar schemes of large-scale lending to many small borrowers in Kenya, Malawi, Sierra Leone, Burkina Faso and Ethiopia. These also could be utilised for providing funds for RETs.

61

Semi-formal sector

There are a number of financial intermediaries that can be termed 'semi-formal'. This means that they are regulated (either by the government or the market), they are highly structured in their organisation, and they fulfil an important role in the financial operations of the country.

Examples of semi-formal lending organisations include savings and credit co-operatives and credit unions, though in some countries these may be classified as the formal sector (Germidis, 1991). In 1989 it was estimated that there were over 17,000 credit unions spread throughout 67 developing countries. These had approximately 9 million members, accumulated savings of $1.8 billion, and outstanding loans of some $1.4 billion (Otero and Rhyne, 1994).

The common characteristics of financial organisations in the semi-formal sector include:

- ability to provide credit otherwise not available
- in-depth knowledge of local clients and inter-linkage between members
- simple transaction bureaucracy
- semi-flexible lending structure
- voluntary participation
- in many cases, there are accumulated savings which allow participation in the lending programme.

Informal sector

In many countries, there is a highly developed informal financial sector. Financial organisations classified as semi-formal in one country may be classified as informal in another, solely due to regulations for lending. The characteristics that these diverse organisations share include:

- provision of credit otherwise not available through the formal sector

Small wind generators can be purchased by cash (e.g. in Mongolia) or credit
Photo: Marlec

- 'closed circuit' knowledge of local clients
- low / no transaction bureaucracy
- flexible lending
- voluntary participation

However, this is where the similarities stop, as the organisations vary from private financial companies (such as those providing hire purchasing and leasing arrangements) to pawnshops and money lenders.

7.2 Bank funds

There are examples in developing countries where commercial banks (both private and public) are lending to small borrowers. However, this is the exception rather than the rule as generally large numbers of small borrowers and/or large numbers of savers with multiple transactions per month are not economical for large, bureaucratic organisations. This is one of the reasons for a large informal lending sector in many countries, consisting of organisations with smaller overheads and a greater ability to provide services for smaller borrowers and lenders.

One way to overcome this problem for the commercial banks is for them to lend to smaller funds, who then on-lend for small loans (such as the cost of a SHS). Such funds can be used in a number of ways:

- direct financing of loans to small borrowers
- lending to the fund itself
- interaction with the fund to provide a joint savings and credit facility

There are benefits and disadvantages in all of these, for both the bank and the fund.

In Figure 7.1a, a scheme is shown whereby a bank lends to a fund. The fund then on-lends to its clients. The bank has a guarantee scheme with the fund to cover its risks. Figure 7.1b shows the costs and income that will be experienced by the fund, and the financial and institutional balances that can be achieved. In this model, the fund does not have a savings facility, but may require potential borrowers to have a savings history with another financial organisation, such as a commercial bank. In many cases, the most appropriate model will be for the fund to broker savings with a bank. This provides the bank with new clients and provides an interface between the bank and people who may not have been in contact with a bank before. It can earn the fund additional revenue, either through access to funds with a lower interest rate or by receiving a fee or percentage for each new customer. The ability to negotiate with a bank is vital for this type of fund.

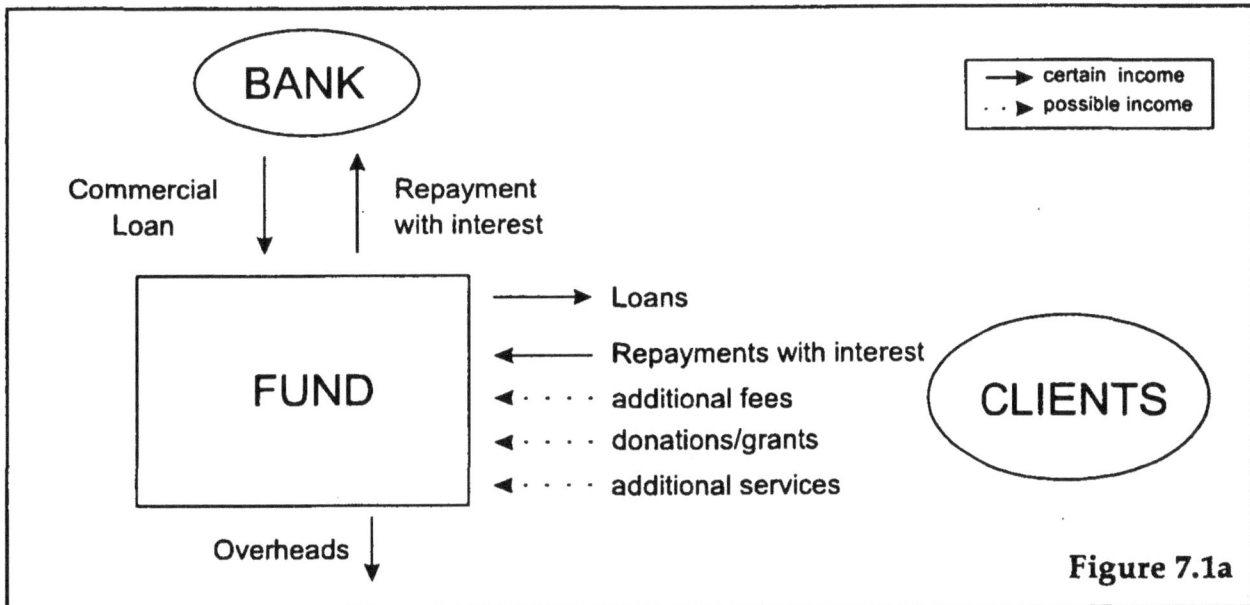

Figure 7.1a

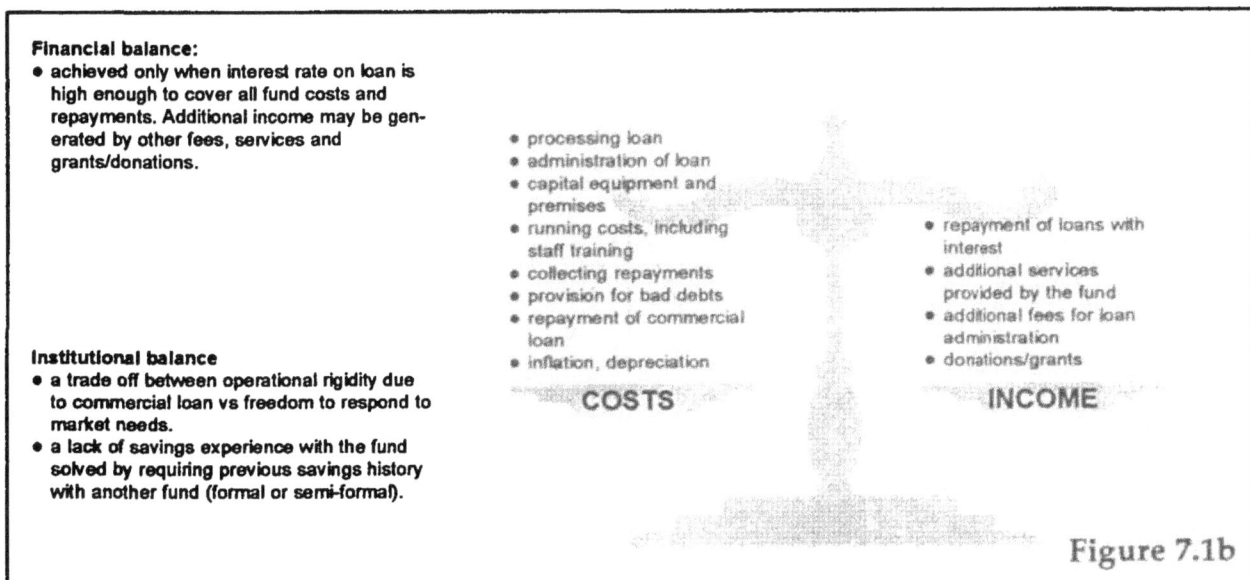

Financial balance:
- achieved only when interest rate on loan is high enough to cover all fund costs and repayments. Additional income may be generated by other fees, services and grants/donations.

Institutional balance
- a trade off between operational rigidity due to commercial loan vs freedom to respond to market needs.
- a lack of savings experience with the fund solved by requiring previous savings history with another fund (formal or semi-formal).

COSTS
- processing loan
- administration of loan
- capital equipment and premises
- running costs, including staff training
- collecting repayments
- provision for bad debts
- repayment of commercial loan
- inflation, depreciation

INCOME
- repayment of loans with interest
- additional services provided by the fund
- additional fees for loan administration
- donations/grants

Figure 7.1b

Figure 7.2a illustrates the interaction between a commercial bank and a fund, where there is a credit facility provided by the bank to the fund. The fund requires its borrowers to save with this commercial bank, as the fund itself does not have a savings facility. In requiring ongoing savings, the fund reduces the risk of default on its loans, as any default can be covered by the borrowers' savings at the bank. In many cases, this increased security means that the commercial bank can lend to the fund at a lower interest rate, as there is less risk to cover. Figure 7.2b shows the financial and institutional balances in this model.

Figure 7.3a illustrates the scenario whereby the fund has its own direct savings and credit facilities. In this situation, it has more savers than borrowers, and more financial and management flexibility. Savings can be utilised for loans, thus lessening the requirement for borrowing from other financial institutions. However, a sophisticated management system must be in place to make sure that the fund is run correctly and makes enough return on its lending to finance its own management, return interest to savers and to cover possible defaults. Figure 7.3b shows the financial and institutional balances.

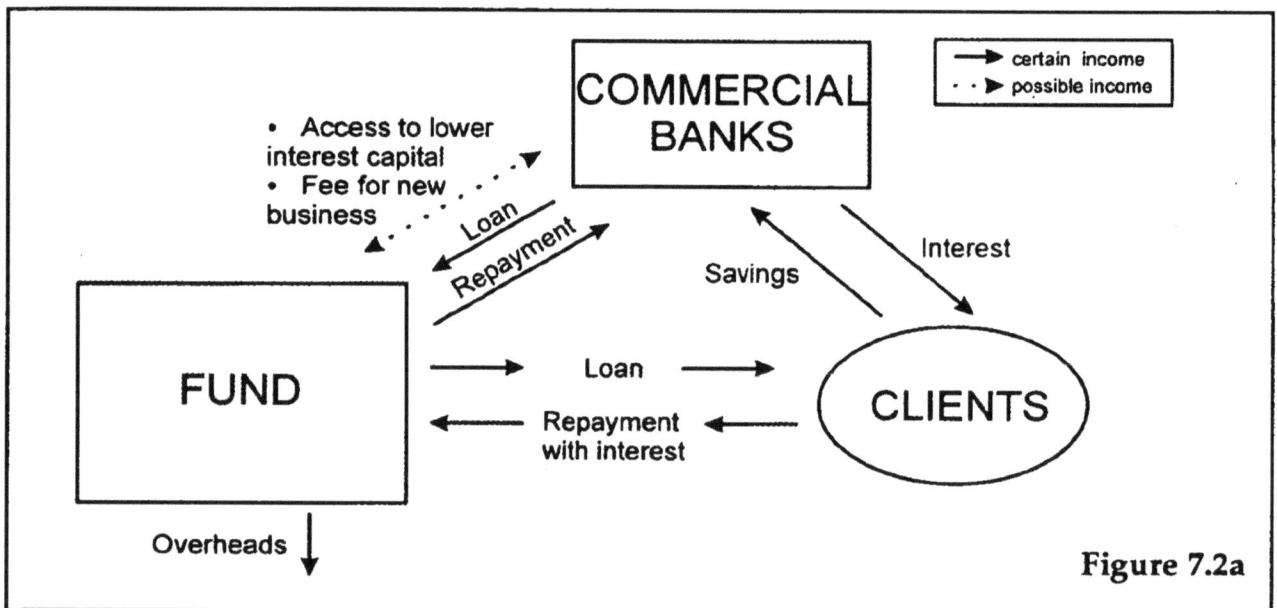

Figure 7.2a

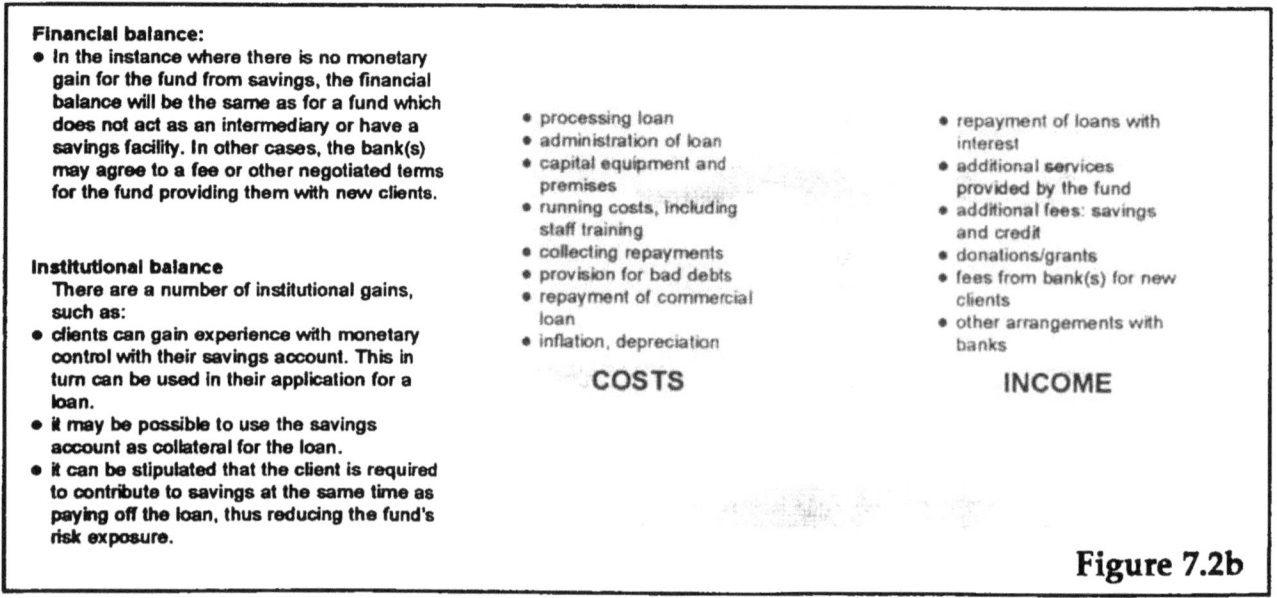

Financial balance:
- In the instance where there is no monetary gain for the fund from savings, the financial balance will be the same as for a fund which does not act as an intermediary or have a savings facility. In other cases, the bank(s) may agree to a fee or other negotiated terms for the fund providing them with new clients.

Institutional balance
  There are a number of institutional gains, such as:
- clients can gain experience with monetary control with their savings account. This in turn can be used in their application for a loan.
- it may be possible to use the savings account as collateral for the loan.
- it can be stipulated that the client is required to contribute to savings at the same time as paying off the loan, thus reducing the fund's risk exposure.

- processing loan
- administration of loan
- capital equipment and premises
- running costs, including staff training
- collecting repayments
- provision for bad debts
- repayment of commercial loan
- inflation, depreciation

COSTS

- repayment of loans with interest
- additional services provided by the fund
- additional fees: savings and credit
- donations/grants
- fees from bank(s) for new clients
- other arrangements with banks

INCOME

Figure 7.2b

The following two bank examples illustrate a number of these important points.

The Grameen Bank of Bangladesh began as an experimental project in 1976. It now has over 7,000 employees, 78% of whom are located in regional or branch offices. In 1988 its portfolio averaged $15.23 million. Its policy is to lend small amounts at a local level, with branch and regional offices being the loan decision-making centres (Otero and Rhyne, 1994).

The Grameen Bank has low profitability. Whilst this may inhibit its flexibility as an organisation, it does not impede its growth or daily operations. It also draws directly and indirectly on donor resources, which are then invested to cover operational costs. From 1985 to 1987 money from donor funding earned interest of around 13 - 17% for the bank, and was an important part of its financial strategy. In 1996, the Grameen Trusts began a programme to lend for SHSs (Selling Solar, 1996).

In Indonesia, a programme of rural credit was implemented in 1984. The programme, which charged commercial rates of interest, was administered by the state-owned com-

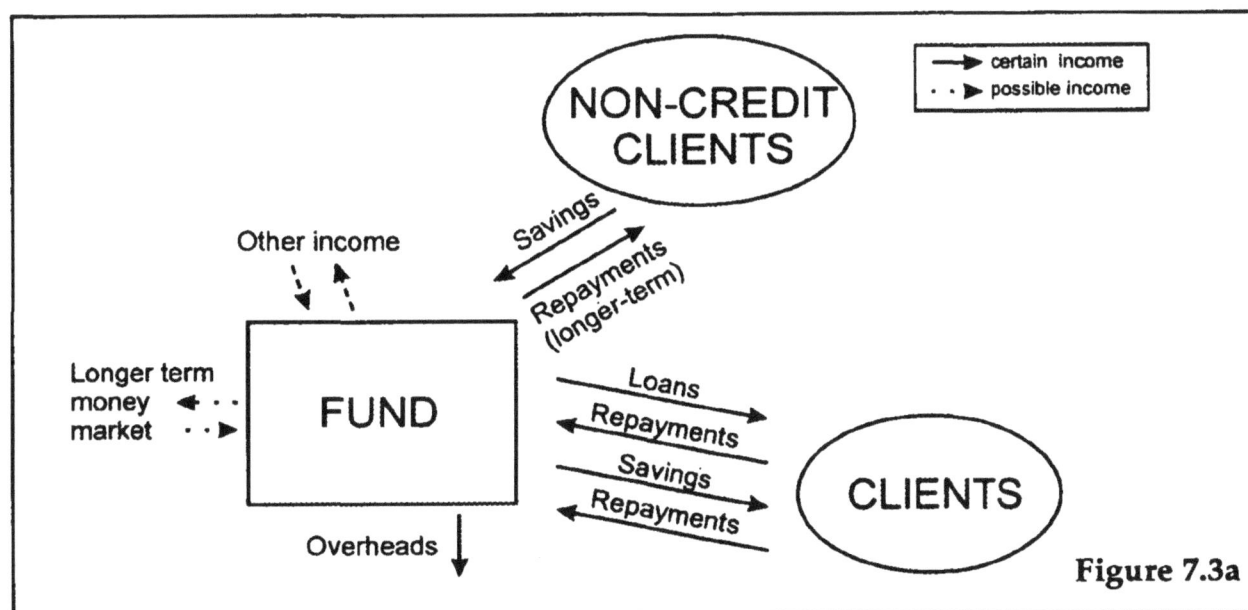

Figure 7.3a

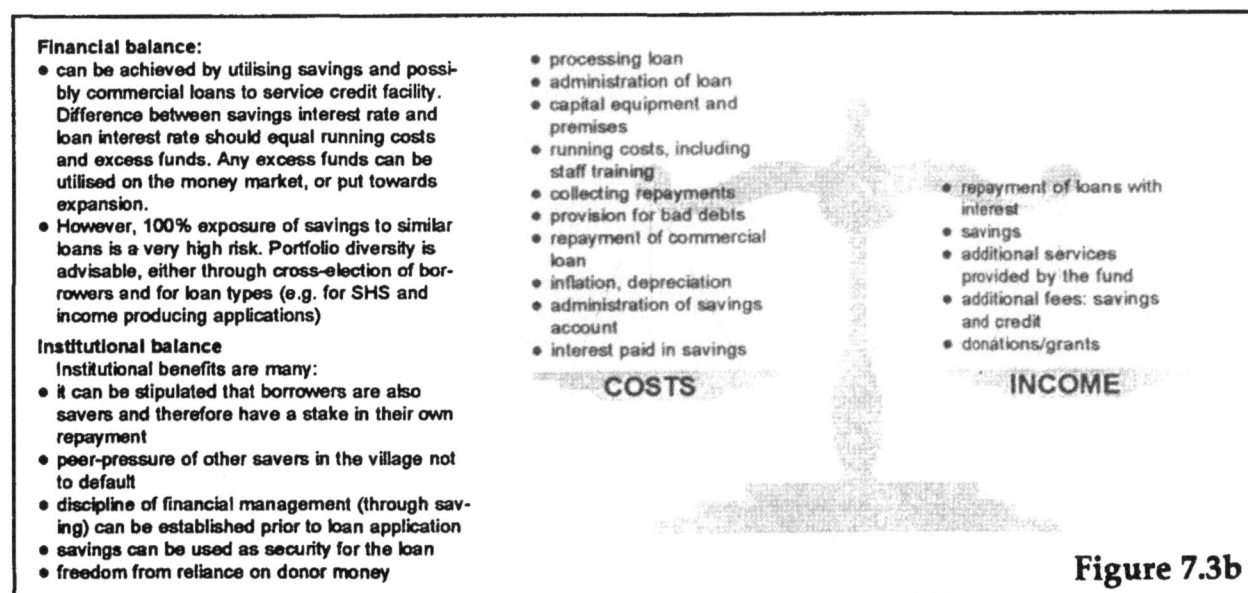

Financial balance:
- can be achieved by utilising savings and possibly commercial loans to service credit facility. Difference between savings interest rate and loan interest rate should equal running costs and excess funds. Any excess funds can be utilised on the money market, or put towards expansion.
- However, 100% exposure of savings to similar loans is a very high risk. Portfolio diversity is advisable, either through cross-election of borrowers and for loan types (e.g. for SHS and income producing applications)

Institutional balance
Institutional benefits are many:
- it can be stipulated that borrowers are also savers and therefore have a stake in their own repayment
- peer-pressure of other savers in the village not to default
- discipline of financial management (through saving) can be established prior to loan application
- savings can be used as security for the loan
- freedom from reliance on donor money

- processing loan
- administration of loan
- capital equipment and premises
- running costs, including staff training
- collecting repayments
- provision for bad debts
- repayment of commercial loan
- inflation, depreciation
- administration of savings account
- interest paid in savings

COSTS

- repayment of loans with interest
- savings
- additional services provided by the fund
- additional fees: savings and credit
- donations/grants

INCOME

Figure 7.3b

mercial bank, Bank Rakyat Indonesia (BRI) and was known as Kredit Umum Pedesaan (KUPEDES). By 1986, rural savings and credit through the KUPEDES programme had become profitable for both the government and the bank, with villagers' deposits increasing from $17.6 million in 1983 to $1.3 billion in 1991. Deposits now cover outstanding loans by around five to one (Otero and Rhyne, 1994; Banking with the Poor, 1992).

Solar Home System, Indonesia   Photo: Novem

What these two examples illustrate is that lending at a local level can be successful and profitable for both commercial lenders and the borrower. Both these examples are primarily for lending to micro-entrepreneurs and for agricultural purchases. There are certain important commonalities between the two programmes, which can be transferred to other funds and programmes:

- lending coupled with savings fulfilled a need at a local level

- there was mobilisation of local financial resources, which reduced the need for external financing

- attention was paid to what the people wanted at a local level and which financial services would best fit these needs - before the funds were established

- profitability was reached and sustained

- peer lending was important for low default rates

- staff were trained to understand the markets that they were financing, including the inter-linkages of local social and economic issues as well as technical aspects

- banking packages were designed to be as simple and as expedient as possible

In the case of KUPEDES, market research undertaken prior to implementing the project

concluded that there was a huge potential for an appropriate large-scale rural banking system. These conclusions turned out to be correct, and vindicated the government's decision and exposure to financial risk.

The Grameen Bank spreads its risk across the international donors, as well as the central government of Bangladesh.

7.3 Client savings

Savings can be a key component of successful credit programmes (Banking with the Poor, 1992; Padmanabhan, 1988). It is particularly useful for a new fund to have a savings component in areas where there is little existing access to savings mechanisms, either formal or informal. Within a savings and credit programme, there may be many savers who do not wish to borrow from the fund and therefore provide it with a steady source of lending capital. Savings programmes are necessary for economic development at a local, regional and national level, by making funds available for lending and reducing the need for foreign capital.

Traditional savings and credit mechanisms already exist in many countries, often in the informal sector. For instance, in Ghana, traditional 'susus' regularly collected funds from members, who then had access to loans from the group pool. In 1989, the government formalised this financial activity by requiring all susus to have an operating licence issued by the Central Bank. This reduced the amount of savings activity initially, but also made the susus more accountable (WWB, 1993).

Many funds require compulsory savings, either with them or an affiliated organisation. For instance, the Grameen Bank requires borrowers to save an established percentage of the loan, which remains as deposit while they borrow from the programme (Otero and Rhyne, 1994). There are a number of advantages of compulsory savings, either before or during a loan period. These include:

- savings can provide the initial link with the banking system, educating the potential borrower and preparing them for the requirements of borrowing from a bank

- savings establish the discipline of regular payments, and can be used as a track record for lending. Savings patterns can be an effective method of client screening for lenders

- if a borrower has savings linked to their borrowings, then it is unlikely that he/she will risk them by defaulting. If this situation or late payment does occur, then the savings can often be used to redeem the default (depending on legalities)

In some countries the law forbids some financial organisations (typically those in the semi-formal and informal sectors) from conducting savings transactions. This can be overcome in a variety of ways, such as those referred to in Figures 7.2 and 7.3.

7.4 Managed funds

These are agreements with other NGOs and can take a variety of forms. For instance, one or a number of NGOs may have money available for lending, but may not have the expertise or may not want to manage the credit programme. The organisation will then contract with other NGOs to undertake the management of the fund on their behalf.

7.5 Grants and soft loans

As shown in chapter 5, grants and loans may be useful in seeding the fund. However, they may be seen by the borrowers as 'free' money: for instance, in the German government's SHS programme in the islands of the Philippines in the 1980s, the SHSs were heavily subsidised. Collection of repayments was also very spasmodic, and the local people came to regard their SHSs as a 'gift' from the 'wealthy' German government and defaulted on their loans (Biermann et al, 1995).

If the loans are 'soft' (that is at preferential borrowing rates), then this will assist the fund in buying down the cost of lending without it being seen as a direct subsidy or gift.

There are many issues that need to be resolved when using grants or soft loans, such as the longevity of the fund. Should early lenders have low interest rates, but those taking out a loan later in the programme (when the cheap finance has gone) have to pay higher rates?

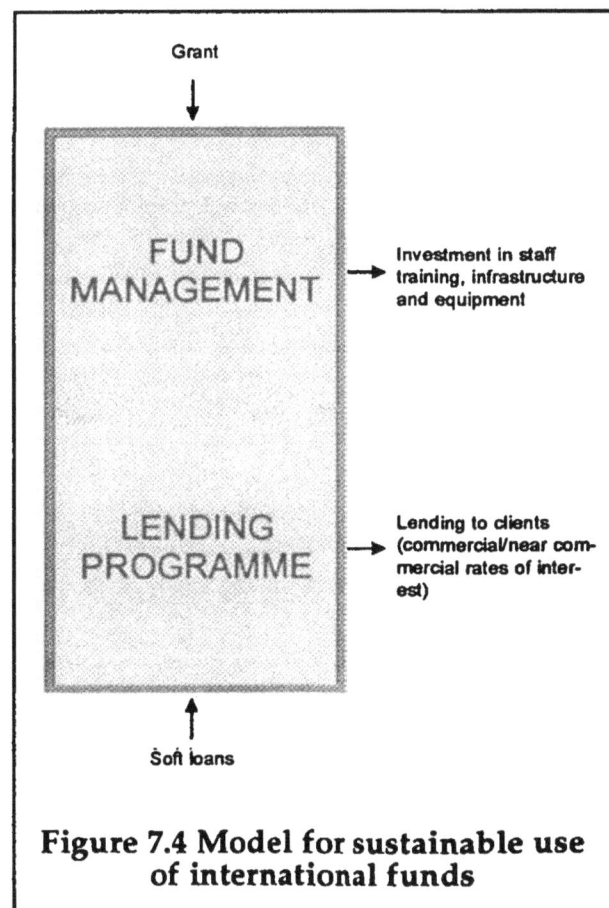

Figure 7.4 Model for sustainable use of international funds

One way to avoid this is to have commercial or near commercial rates of interest throughout the lifetime of the fund. As suggested previously, any additional income can then be used to invest in staff development, expansion of the fund and increased services to members.

Instead of acting as a direct subsidy to a programme or fund, grants in particular can be a good source of funding for staff development and capital equipment. When used in such a way, the grant does not affect the lending capital as such; it does not create lower interest rates (which then have to be raised later on to cover the cost of inflation and defaults), and it becomes a long term investment in both local people and the Fund. Figure 7.4 shows this concept diagrammatically. Any grant or seed money is used for the management of the fund (i.e. investment in staff, training, infrastructure development and equipment), whilst loans are used to capitalise the fund lending for RETs.

There are a number of financing schemes in operation today for lending for SHSs in particular. These involve a variety of organisations: government departments, the utility, NGOs, banks and/or international agencies. They all have a technical component, that is,

they have made adequate provision for installation, maintenance and repair of their systems. The type of scheme varies, reflecting the differences in local conditions for financing. All have been assisted by international agencies, though the source has varied and includes the World Bank, the Global Environment Facility, international funds (such as Rockefeller and other foundations), bilateral aid and utility funding. Chapter 8 discusses the implementation issues for subsidies and grants in more detail.

7.6 Conclusions

There are a number of avenues open for RETs to source funding. Each of these has advantages and disadvantages, depending on local conditions. However, there are a number of commonalities for successful sourcing and lending of funds. Table 7.1 lists different types of financial organisations, summarises their ability to support RETs financing (direct and indirect), the type of finance which they could provide (e.g. equity finance, trade finance etc.), lists any examples of PV financing to date, and summarises the advantages and disadvantages of using such organisations and their schemes.

SOURCE OF FUNDS

Table 7.1: Financial organisations and infrastructures in developing countries

A. FORMAL SECTOR					
Type of organisation	Appropriate to RETs financing	Type of finance relevant to PV	Advantages	Disadvantages	Known PV examples to date
Central Bank	Indirect support only: dispersion of funds to intermediaries; co-sponsored intermediary funds	Equity capital for financial markets Co-sponsoring special funds for lending, e.g. Credit Guarantee Schemes	• can assess international funds at preferential rates	• usually bureaucratic structure • centralised • high lending costs for small borrowers	• Bank of Ceylon, Sri Lanka
Commercial Banks	Direct support:RETs systems Indirect support: channelling funds to other intermediaries possible co-partnerships with ROSCAs or co-operatives	Loans for PV systems (small and large)	• developed lending and savings functions • can have wide geographical spread • access to international finance markets	• favour large-scale borrowers in urban areas • high transaction costs • high transaction bureaucracy • require high collateral security on loans • high interest rates • stringent collateral requirements	• The People's Bank, Sri Lanka • Bank Rakyat, Indonesia • Caisse Nacionale de Credit Agricole, Morocco • Syndicate Bank, India
Merchant Banks	Support to companies only	Trade finance for PV companies (local and export)	• access to international finance markets	• *Same as for Commercial Banks*	Many, world-wide for trade finance
Savings Banks	Indirect support: channelling funds to other intermediaries	Savings performance can be used as a reference for lending	• established networks • savings and credit facilities	• urban bias • little/no knowledge about PV to date • transaction bureaucracy • stringent collateral requirements	• Banque du Rwanda
Development Banks	Direct and indirect support	Act as intermediary for loans dispersal	• long term financing • history of financing new industry • technical services • accept deposits from institutions • below market interest rates	• generally do not accept deposits • prefer large borrowers and industry • high level of transaction bureaucracy • stringent collateral requirements	• Grameen Bank • Banco Popular, Dominican Republic • Banco del Nordeste do Brazil

69

FINANCING RENEWABLE ENERGY PROJECTS

Type of organisation	Appropriate to RETs financing	Type of finance relevant to PV	Advantages	Disadvantages	Known PV examples to date
Building Societies	Direct support for SHS and other building-related products	Finance for housing-related loans	• savings and loan facilities • can have wide geographical spread • close to the local population • low loan interest rates • low level of transaction bureaucracy • do not have stringent collateral requirements	• may be limited by own regulations for lending for SHS • prefer public enterprises and co-operative lending • low interest rates on savings	Not known
Postal savings networks	No	Savings only. However, could be linked to another fund which does not have savings facilities	• established networks with local populations • well known	• minimum savings amounts • no loans	

B. SEMI FORMAL SECTOR

Type of organisation	Appropriate to RETs financing	Type of finance relevant to PV	Advantages	Disadvantages	Known PV examples to date
Fixed fund associations	Direct loans for RETs systems	PV systems community loans	• do not have stringent collateral requirements	• sometimes only savings • reliant on honesty of their officers • usually short term loans	Not known
Rotating Savings and Credit Associations (ROSCAs)	Direct loans for RETs systems	PV systems community loans	• can be regulated • well established in many countries and in rural areas • savings and loans • do not have stringent collateral requirements	• dependent on honesty of ROSCA officers • officers may receive little financial and admin training	Not known
Credit co-operatives (private or public)	Direct loans for RETs systems	PV systems community loans	• can be linked into other organisations • can be regulated • well-established in many countries and in rural areas • savings and loans • can be linked through formal sector • do not have stringent collateral requirements	• susceptible to local economic disasters • no interest on savings	• BANPRES, Indonesia • Sarvodaya, Sri Lanka • Vietnam Women's Union • Rural Co-operatives, Bolivia

SOURCE OF FUNDS

Type of organisation	Appropriate to RETs financing	Type of finance relevant to PV	Advantages	Disadvantages	Known PV examples to date
Village banks	Direct loans for RETs systems	PV systems community loans	• per group lending • savings and loans • history of international seed capital • can be linked to the formal sector • no stringent collateral requirements • members receive share of profits	• fairly new financial phenomenon (1980s) • no interest on savings • limited access to capital • dependent upon strong local leaders	Not known

C. INFORMAL SECTOR

Type of organisation	Appropriate to RETs financing	Type of finance relevant to PV	Advantages	Disadvantages	Known PV examples to date
Money lenders (commercial and non-commercial)	Possible	PV systems	• immediate credit	• high interest rates • short term loans	Not known
Partnership firms (e.g. Chit Funds in India)	Possible	Usually finance trade, transport, industrial and export activities	• regulated (in India) • often have interest rate ceilings • locally based but widespread • often linked to commercial banks	• 'auction' system for funds	Not known
Pawnbrokers	Possible	Loans for SHS	• sometimes regulated (e.g. India) • flexible collateral requirements • easy valuation	• high interest rates • usually short term	Not known
Finance corporations; Partnership firms / Private / Public limited companies	Possible	PV systems	• can be regulated (e.g. India) • do not have stringent collateral requirements	• high interest rates • usually short term repayment schedule	Kwazulu Finance & Investment, South Africa
Hire purchase companies	Direct financing of RETs systems	PV systems	• schemes may be already known through financing of consumer durable goods (e.g. sewing machines) • do not have stringent collateral requirements	• Usually short term repayment schedule • high interest rates	Singer Corp., Sri Lanka

FINANCING RENEWABLE ENERGY PROJECTS

Type of organisation	Appropriate to RETs financing	Type of finance relevant to PV	Advantages	Disadvantages	Known PV examples to date
Leasing companies	Direct financing of RETs systems	PV systems	• able to resource own financing • able to provide complete service	• may be more expensive than other options	Not known
NGOs	Direct financing and facilitation of fund disbursement	PV systems; community systems	• possible access to international development funds • flexible lending	• many NGOs do not have the 'corporate climate' to lend • government regulations may forbid NGOs to have a savings function • often a lack of formal training in lending and fund management	• Enersol, Dominican Republic • Solar Electric Light Fund • Solanka, Sri Lanka • ADESOL, Dominican Republic • Gansu SELF, China • Servicios Sociales de la Iglesia, Dominican Rep.
Energy Service Companies, ESCO (e.g. NGO, utility, private company)	Direct financing of PV systems	PV systems; community systems	• provide complete service including maintenance • can be a variety of organisational structures • long term loans can provide large community systems as well as small systems • possible to access equipment through international tenders	• the ESCO owns the system • a wide range of services are required from one organisation • can have higher organisational running costs	• Utility - Tuvalu Solar Electric Co-operative Society • Co-operative - Rural electricity Co-ops, Philippines • Private - Soluz, Dominican Republic

72

Implementing a Revolving Fund

8

Earlier chapters have presented various appropriate mechanisms to help make RETs more affordable for many people in developing countries, and introduced the revolving fund concept. This chapter looks at the revolving fund in more depth and illustrates the factors which can contribute to the success or failure of such a system.

In practice, there are no standard blueprints which development workers can follow to guarantee the performance of their chosen mechanism for financing RET deployment. There are many parameters which need to be accounted for, and each case must be adapted to the specific local environment. This chapter presents an outline of the key factors which can influence the establishment and performance of a 'typical' revolving fund mechanism. Many of the considerations in this example are directly applicable to other credit disbursement mechanisms.

Although exact programme details will vary from one case to the next, the underlying concept is unchanging: in basic terms, a balance must be struck between the unit price of credit offered to a customer and the unit cost to the programme of providing that credit.

8.1 Modelling a revolving fund

This Guide employs a spreadsheet-based financing model to illustrate some of the factors which might affect the performance of a revolving fund. The model simulates an organisation that sells or rents out PV home

systems in a developing country. The model organisation does not slot directly into any one of the various arrangements for financing RETs listed earlier (lease, service or rental), but instead crosses the boundaries of all three. By adjusting the input parameters it is possible to demonstrate how a revolving fund might be applied in each case. A more detailed explanation of the model concept is given in Appendix 2.

Credit disbursement to enable system sales on finance and/or rentals requires the securement and management of capital which, in this case, is achieved through a revolving fund. Revenues into the revolving fund are counterbalanced by the costs of establishing and managing the fund. The difference between the two effectively determines the number of systems that can be financed.

A continuous negative annual net cash-flow (i.e. income is less than expenditure) would be a drain on the overall fund unless a continuous external subsidy was available to balance the loss. Temporarily negative net cash-flow need not be a significant cause for worry, however, providing either the net loss is very small, or the short to medium-term forecast indicates a positive balance. This explanation will be clarified in the examples that follow.

Finally, although the model uses PV solar home systems for illustration purposes, it could be adapted to illustrate sales of other RETs or even how revolving funds might relate to other micro-enterprise ventures.

Covering costs

As the historical lessons above illustrate, one of the major factors governing the success or failure of the revolving fund concept is the extent to which the costs of providing the credit service are covered.

A project will have two principal sources of funds: initial funding - be that from a seed grant or debt finance - and customer payments. These inflows are used to purchase system hardware, cover installation and possibly some or all service operating costs. The contribution of customer resources will not only affect user charges, and hence the prospective target group, but also the level of sustainability of the project, as this is a 'guaranteed' continuous cash flow, required to provide technical and credit servicing and project expansion.

Credit interest charges should at the very least be sufficient to cover the costs of providing the service. Service costs consist of three discrete components:

- costs of obtaining and maintaining the resources which are on-lent to borrowers. Included in this are interest charges and repayments of debt capital, interest or dividends payable on any savings which might be used as a source of funds, ensuring protection against inflationary pressures and against possible devaluation of local currency when debt repayments are in a stronger currency

- programme operating costs. These include payment of staff salaries, maintaining an office, loan processing, disbursement and repayment collections, equipment purchase/depreciation, system service and maintenance (where appropriate), materials purchase and transport

- lending risk costs. Such costs are associated with possible bad debts and delayed payments

Demonstrating a RETs system before beginning a programme is very important for gaining people's confidence in the technology and stimulating demand

Photo: BP Solar

Aside from accessing least-cost finance, the first component is largely beyond the control of the project management, but the other two can be influenced. Improvements in operations through more efficient loan processing and collection procedures, reduced maintenance commitments and better assessment and control of repayment risks can reduce service costs, thereby reducing customer charges without damaging the fund.

If repayments are not geared to cover minimum costs of providing the service, then the fund will be reliant on continued external support and will, in the longer term, be unsustainable.

If there is sufficient demand for the product or service, and repayment levels are such that all the costs of operating the fund are covered, then the fund will be sustainable. If repayments also allow for a slight 'profit' the fund may grow exponentially.

The most sustainable form of revolving fund is organised so that it is able to:

- provide loans at market prices

- provide for its own administration

- provide for the maintenance and installation of systems which it is financing

74

- expand its client base steadily

- expand/maintain adequate lending capital

Unfortunately, past experiences have demonstrated the ease with which some costs (which can be a significant drain on available resources) can be underestimated or overlooked (IT Power & Eurosolar, 1996). It is therefore essential, prior to initiating any financing programme, be it based on a revolving fund or other mechanism, that a full and thorough assessment be undertaken of the costs involved.

Obviously it is not always possible to guarantee 100% accuracy where cost estimates are involved, but effective groundwork at this stage can minimise the possibility of damaging fund depletion once the programme is underway. It is also good policy to establish a contingency reserve which could be called upon to cover any unforeseeable expenses. This could be either a fixed annual amount or a percentage of the annual sales.

Some of the major costs which could conceivably be incurred in establishing and operating a revolving fund are detailed in Table 8.1. The influences of various costs on the project management costs and overall performance of the revolving fund are illustrated in examples 2 and 3 later on in this chapter. The 'balance' to these numerous programme costs is mainly the repayments from customers. Repayment annuities are dependent on the initial amount borrowed - in this case the cost of the system being purchased inclusive of any import or sales-tax plus any mark-up (the mark-up is effectively the 'profit' which the vendor would make if they sold the system for cash) - the period of the loan and the interest rate charged. The interest rate is the fund-managers tool for covering these costs and ensuring that the fund will not be eroded (see Table 8.2).

The right seed

Revolving funds can be established with grants or debt. The choice or availability of seed capital can have a significant impact on the cost of operating the fund.

For instance, if seed funding is a grant (i.e. does not have to be repaid) then the overall cost of the fund will be significantly lower than if the same seed is financed with debt (i.e. a loan). If the revolving fund is to be sustainable, credit repayments will not only have to cover the costs of depreciation of the original fund but will also have to be sufficient to meet the repayments of the original seed loan. If interest is also charged on the seed capital, the additional cost will be transferred to the borrower through higher repayments.

Different project aims can have a direct bearing on the sources of financing available to the project, which in turn will have an impact on the projects costs. For instance, where the primary concern is improvement in living

---

Table 8.1 Financing programme costs

Start-up costs
(one-off charges at financing programme initialisation):
- purchase of land / buildings
- purchase of office equipment
- purchase of vehicles

Recurring fixed costs
(independent of number of sales/rentals):
- salaries at HQ (e.g. of management, marketing and administrative staff)
- marketing activities
- building / equipment rentals
- utility bills
- office supplies
- interest on outstanding debts
- depreciation of fixed assets

Recurring variable costs
(dependent on number of sales/rentals):
- materials (e.g. system bulk purchases, including import duty / sales tax where applicable)
- component replacements
- vehicles (fuel requirements thereof plus any new purchases)
- salaries of field staff
- cost of delayed or non-payments (usually a percentage of total sales)
- corporation tax payable on profits
- inflation

Table 8.2 System prices and profitability

The total price of systems sold on finance terms is composed of:

	System Cost	(largely fixed)
+	*Import/sales tax*	(fixed)
+	*Mark-up*	(controllable)
+	*Total Interest*	(controllable)
=	*System Price*	

The duty payable is a cost to the fund. The mark-up on the system cost, and even more importantly, the interest charged for providing the credit service, are therefore the key tools with which the programme management organisation can exert some control on the performance of the fund.

N.B. The total price of a system may vary depending on how long the customer takes to repay the loan. The total interest charged will increase if the repayment period increases.

The profit (or loss) made on each system sold is calculated as:

	System Price
-	*Marginal System Cost*
-	*(Fixed Cost / No. of Systems Sold)*
=	*Profit (or loss)*

As the number of systems sold increases, the contribution of the fixed cost component on the programme's profits (or losses) is reduced.

standards, including income generation, educational or environmental activities, grants or low-interest debt financing may be available from local or national governments. International and national charities are also potential sources of funds for such projects.

The influence of seed fund sourcing on the operation of the fund is highlighted in the examples that follow.

Mitigating the damage from bad debts

Non-payments can be a significant cause for concern for fund managers. Although some bad debts are attributable to delinquency, this is not always the case. Failure to meet the repayment schedule may arise, for example, as a result of failure of a borrower's main source of income. Farmers, for instance, will be very badly affected if harvests are poor.

While bad debts arising through loss of income are unpredictable and generally unavoidable, it is possible to limit the potential damage to the lending programme by ensuring that the lending portfolio - in terms of the target customers and even the activities that the fund will finance - is well balanced.

This also demonstrates a situation when the contingency reserve fund mentioned earlier could be pressed into action.

The other cause of bad debt (i.e. delinquency) is perhaps more controllable through the early action of the fund management staff. As soon as any signs of repayment trouble are apparent, field staff should be assigned to establish the extent of the problem, to urge timely repayment, and if appropriate to determine a revised repayment schedule. This obviously relies on well-trained and astute field staff.

The value of this approach is illustrated by both the Grameen Bank (Yaron, 1992) and the Women's World-wide Banking organisation (WWB, 1993) which have reported loan collection rates in excess of 97% while lending to some of the most financially disadvantaged clients.

Solidarity groups as mentioned previously rely on peer pressure which helps to reduce repayment problems. The Grameen Bank is one notable initiative which employs a peer pressure mechanism.

Expansion and staff training

Expansion of the programme will often be closely tied to recruitment and training of new staff or retraining of existing personnel. This can be an important cost consideration. However, funding can often be obtained from international grants, as part of human resource development programmes rather than as straight aid, which avoids the imposition of additional project costs that could jeopardise the fund.

What cost to the customer?

To ensure that all project costs are adequately covered and that the fund is not eroded requires that income at least equals expenditure. External subsidies and other income from non-fund-related activities can contribute to overall project sustainability (though reliance on any subsidies obviously implies that the programme is not self-sufficient). Thus, transferring at least some of the project costs to the borrower is realistically unavoidable.

Surprisingly though, efforts aimed at avoiding fund depletion do not necessarily imply a significant added burden to the borrower. Adjusting repayments by a matter of a few dollars a year for every customer by a slight increase in interest rates can have major repercussions on fund performance, but will not necessarily be a very drastic measure for the individual borrower.

As Table 8.3 illustrates, raising interest rates by 0.5% (interest rates in the range of 20-30%) on a loan of $500 increases the annual repayment annuity over a three-year term by less than $2. For monthly repayments the annuity increases by less than $0.14, and weekly this translates to an increase of approximately 3 cents. Taken as a whole, though, if 1,500 customers are making repayments at any one time (equivalent to 500 loans issued each year), raising repayment interest rates from 20 to 20.5% equates to an increase in fund income of almost $2,800 which could be sufficient for instance to pay the annual salaries of two field staff.

Table 8.3 Example of interest rate variation on customers' repayment annuities

Customers' Basic Terms

Interest rate (%)	20
Repay period (years)	3
Amount borrowed ($)	500

	($/year)	($/month)	($/week)
Normal repayments*	237.36	18.58	4.27

If interest is...	Repayments are:		
	($/year)	($/month)	($/week)
20.00%	237.36	18.58	4.27
20.50%	239.22	18.71	4.30
21.00%	241.09	18.84	4.33
21.50%	242.96	18.97	4.36
22.00%	244.83	19.10	4.38
22.50%	246.71	19.22	4.41
30.00%	275.31	21.23	4.87
30.50%	277.25	21.36	4.90
31.00%	279.19	21.50	4.93
31.50%	281.13	21.64	4.96
32.00%	283.08	21.78	5.00
32.50%	285.03	21.92	5.03

Assumes repayments are made at the end of the period

Example 1 - Grant seeded lease organisation

The base case scenario outlined below presents an organisation offering medium term (five year) lease finance for the purchase of PV solar home systems. The figures are intended to be illustrative rather than definitive, although (where possible) appropriate costs and prices have been used (Wijesooriya, 1994).

The credit facility is provided from a revolving fund seeded with a grant of $450,000. Project start-up costs include land, buildings, office equipment and initial transportation purchases. These costs, estimated at $35,000 are met from an additional grant. In both cases debt or equity could conceivably have been used in place of the grant.

Interpreting the charts in the following examples

Chart (a) in the following figures represents the 'bank-balance' of the programme over its operational lifetime. A programme which performs well - i.e. one where on average the income from customers exceeds the cost of operating the programme over each period - will show steady growth in the bank balance as the programme continues. The value of available funds at the end of each period (a period being one year in these examples) fixes the upper limit on the number of systems which may be financed in the next period (i.e. available funds / average loan value = max. no. of disbursements).

The programme cash-flow over each period, chart (b), indicates whether the fund is making or losing money. A negative cash-flow for any period shows that the programme costs outweigh the income (largely from customer payments) during that time. The cash-flow is therefore the difference between the income to and outgoings from the programme which are presented in chart (c).

The cash-flow is perhaps best used as an indicator for the fund manager. If the cash-flow is positive in any period the fund manager should be fairly happy as he/she knows that the fund balance showed an increase over the previous period and further loan disbursements are possible. If cash flow is negative, this should be a warning for the fund-manager: the fund has eroded somewhat since the previous reporting period. If the 'bank-balance' is large compared to the erosion (as might be expected at the beginning of a programme, for instance) this may not be a problem. If in the next period a similar number of loans are disbursed and cash-flow is positive, or if it is still negative but to a lesser degree than the previous period, and the fund balance is still high compared to the erosion, the manager may relax somewhat. The trend would appear to suggest that positive cash-flow should be achieved shortly and the available funds will probably not be completely disbursed in the interim.

However, if cash-flow is increasingly negative in consecutive periods, this is a major cause for concern as it would tend to demonstrate that the income/cost balance is incorrect. Rather than being replenished by customer repayments, the fund is being drained. In such circumstances the fund income - which we have mentioned comes mainly from repayments - would have to rise. In other words, interest charged to the customer would have to be increased.

Chart (d) plots the cumulative number of disbursements (i.e. system installations) achieved over the duration of the programme. If the funds allow, the number of installations can increase year on year as the programme gains maturity and users' acceptability of the technology increases.

In practice, it is probable that only a proportion of the total fund would be pressed into immediate service, particularly until the programme becomes established. In this example, the financed sales prediction for the first year is 300 systems. It is assumed (from then on) that sales will grow at the rate of 10% annually.

The system available for lease purchase comprises a 30 Wp panel and associated components (charge controller, wiring, switches, battery, lamps, support). Some of the system components are imported. These items are subject to a 30% import tax. All components have a 20 year life-span except for the battery (three years) and the fluorescent lamps (four years). The basic system cost, including import duty where applicable, is estimated at $435.

Customers make a downpayment of 20% of the cash sales price. The cash sales price is based on a 20% mark-up on the system cost. Sales are subject to 15% sales tax. Finance terms include interest at 20% per annum which corresponds to a repayment annuity of approximately $160.

Marginal costs relating to installation, payment collection and servicing, where appropriate, are assumed to amount to $35 per annum. This is based on averaged costs of any components which need to be replaced during the lease period (i.e. one new battery after three years and one new lamp after four years), and on the costs associated with a technician visiting the installation to make repayment collections or to service the equipment.

Annual fixed costs amount to $21,000 to cover HQ salaries, utility payments and supplies. The marketing budget is fixed at $10,000 per annum. A fixed contingency cost of $5,000 is also budgeted for.

All other assumptions are detailed in Appendix 2.

The charts in Figure 8.1 plot the progress of the fund, together with details of system installations and cash-flow over a period of 20 years.

During the first four years of programme operation, the costs (mainly attributable to the purchase of new systems) outweigh the revenue received in the form of deposits and repayments [Figure 8.1 (c)]. As a result the net cash-flow over this period is negative [Figure 8.1 (b)]. However, as more customers join the scheme, the annual repayments rise significantly such that by the fifth year (which in this case coincides with the customers' loan repayment period) income has exceeded expenditure.

The fund, which until then had been in decline, now sees a steady growth. This is enhanced by the de-emphasis of the fixed cost component as the number of systems in the scheme increases (as noted earlier).

It is assumed that when the programme expires after twenty years no new customers are financed. However, repayment collections and servicing are still undertaken. This accounts for the sudden positive jump in net cash-flow which gradually diminishes as the scheme is relieved of the last of the participants.

At the very end of the programme, i.e. year 25 in this case, over 17,000 systems have been financed [Figure 8.1 (d)]. The final fund shows a sizeable positive balance and could be used for some other form of community benefit scheme or a similar programme in a different location. In this way, the use and benefits of the financial resources will be maximised.

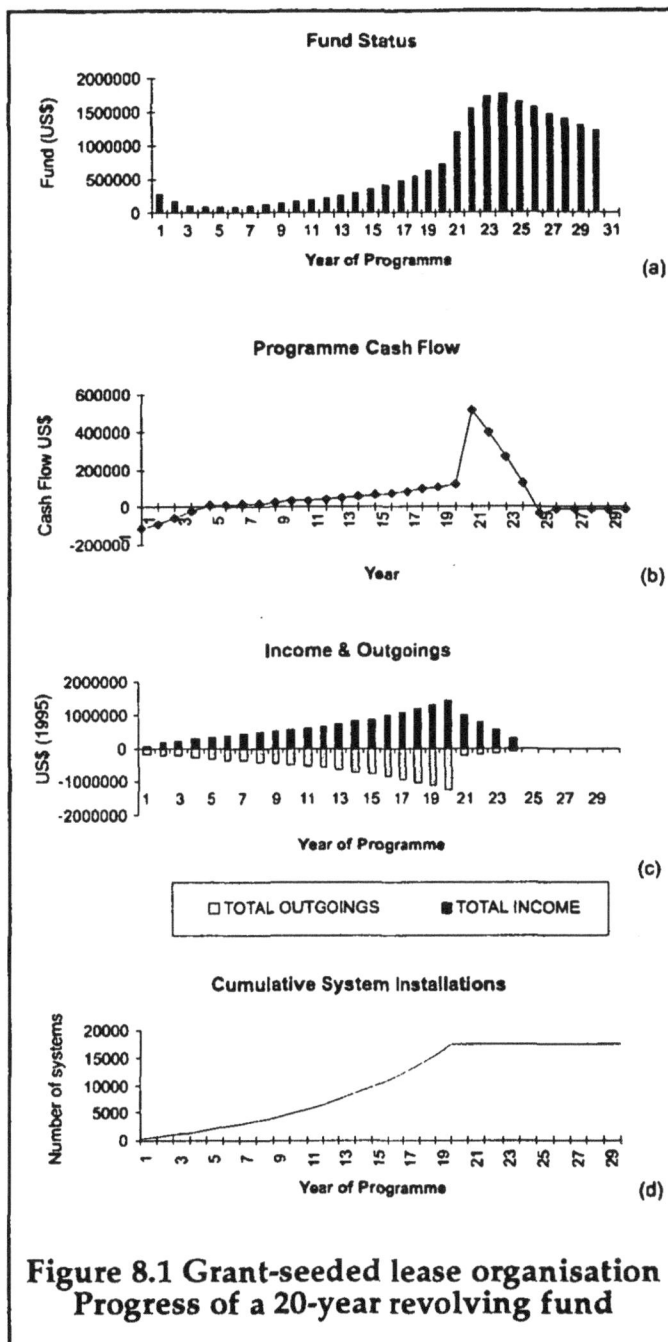

Figure 8.1 Grant-seeded lease organisation Progress of a 20-year revolving fund

Table 8.1 Parameters

Project Finance Summary	
Granted Seed Amount	$ 450,000.00
Debt Seed Amount	$ -
Grace period (yrs)	n/a
Debt repayment period (yrs)	n/a
Debt Interest Rate	n/a
Default Rate	2%
Customers' Finance Terms	
Principal Borrowed	$ 480.24
Repayment period (yrs)	5
Customer Interest Rate	20%
Customers' Annualised repayment	$ 160.58

It is important to appreciate that this is only intended to illustrate the revolving fund concept. In reality, the fund managers may choose to adjust the number of financed sales and payments on a year-by-year basis and aim to have minimal funds remaining at the end of the programme. Ideally, however, the programme should be designed to recover all the costs.

Example 2 - Does debt make a difference?

Intuitively, the answer to this question should be a definite 'yes!' because a project seeded with debt will encounter all the costs of a comparable grant seeded project, and have the additional cost burden of loan repayments and interest payable on that loan.

This example illustrates the possible effect of replacing a granted seed with debt finance. For comparison all parameters and assumptions are identical to the base case example 1, except that the $450,000 grant has been replaced with a loan of the same amount repayable over an eight-year term following a two-year grace period. Interest on the debt is charged at 10% per annum. Ideally interest payable on the outstanding debt would be waived during the grace period, though in practice it is highly probable that interest would be charged.

It is noticeable that in comparison to the base case example the sales profile is poor, but the example does serve to indicate the value of the grace period. Delaying the initial seed debt repayment instalment for two years through the grace period does at least allow the programme to get off the ground. A reasonable number of borrowers are serviced early on which provides some revenue essential to repay the debt. As would be expected, during this period the programme performs similarly to the grant-seeded fund presented in example 1.

As the previous example highlighted, there is a large discrepancy between income and expenditure during the early

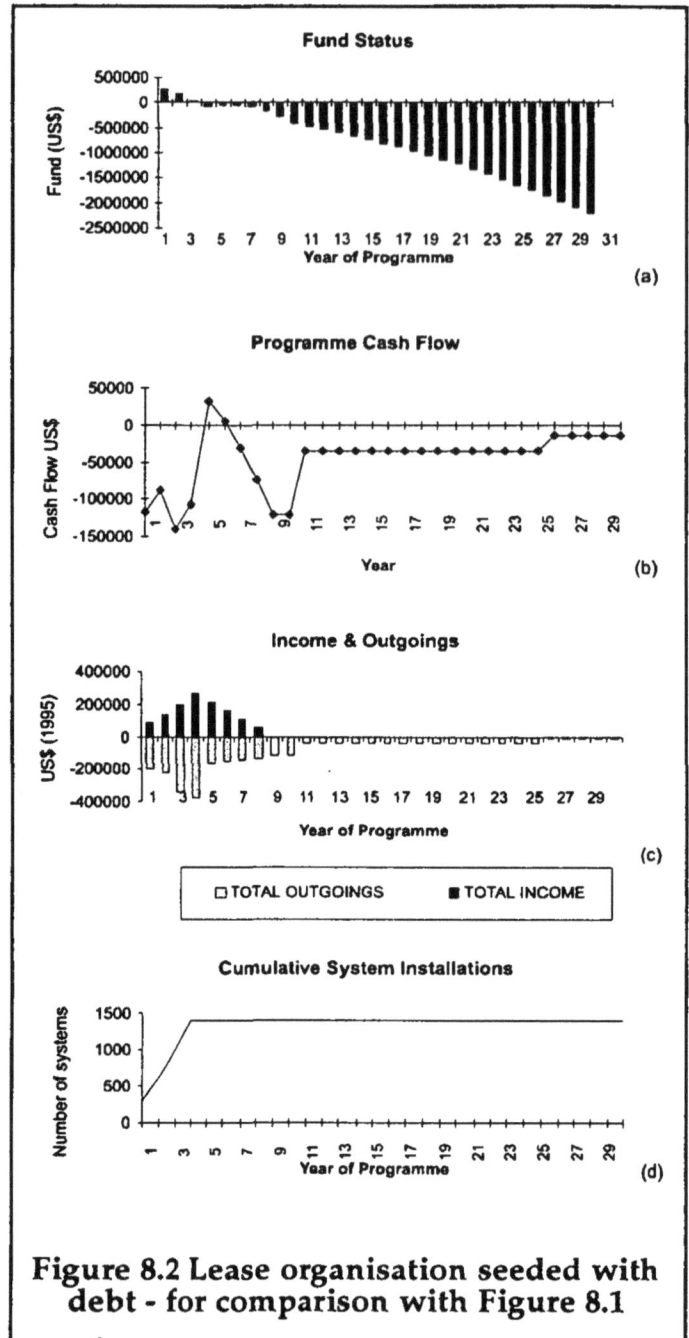

Figure 8.2 Lease organisation seeded with debt - for comparison with Figure 8.1

Table 8.2: Parameters

Project Finance Summary		
Granted Seed Amount	$	-
Debt Seed Amount	$	450,000.00
Grace period (yrs)		2
Debt repayment period (yrs)		8
Debt Interest Rate		10%
Default Rate		2%
Customers' Finance Terms		
Principal Borrowed	$	480.24
Repayment period (yrs)		5
Customer Interest Rate		20%
Customers' Annualised repayment	$	160.58

years of the programme when there are insufficient participants in the scheme to offset the costs of bulk purchase of new systems. When repayments on the seed debt begin, there is a further significant increase in the programme's outgoings.

To make best use of the grace period, it is important to try and achieve a positive cash-flow, or at least minimise the negative cash-flow, as early as possible to reduce the impact of the major debt repayments on the balance of funds. The obvious way of achieving this is by financing a large number of systems in the first year when substantial funds are available and then reducing the number of loan disbursements considerably in the next few years before allowing financed sales to grow again. (This would of course be reliant upon there being sufficient demand for systems.)

In this case, the income from customer repayments and deposits is significantly below the overall expenditure so that there is a large negative net cash-flow, drawing heavily on the revolving fund. The cash flow for year 5 appears to be positive, but this is a false representation that arises because no systems are financed in that year (as there are no funds available) while revenue is being received for all systems that have been installed since the start of the programme. In subsequent years, when borrowers from the early years of the programme have completed their repayment schedule, because no new systems are being financed there is a steady decline in income. Furthermore, because fewer systems are being financed, the fixed costs for each system become much more significant. From then on, it is all downhill and the programme is bound to fail.

This analysis is not intended to imply that debt financed revolving funds are unworkable. The intention is rather to highlight the increased cost aspect which needs to be hurdled and to illustrate that, all other things being equal, the growth of a debt financed revolving fund will initially be far slower than a fund seeded

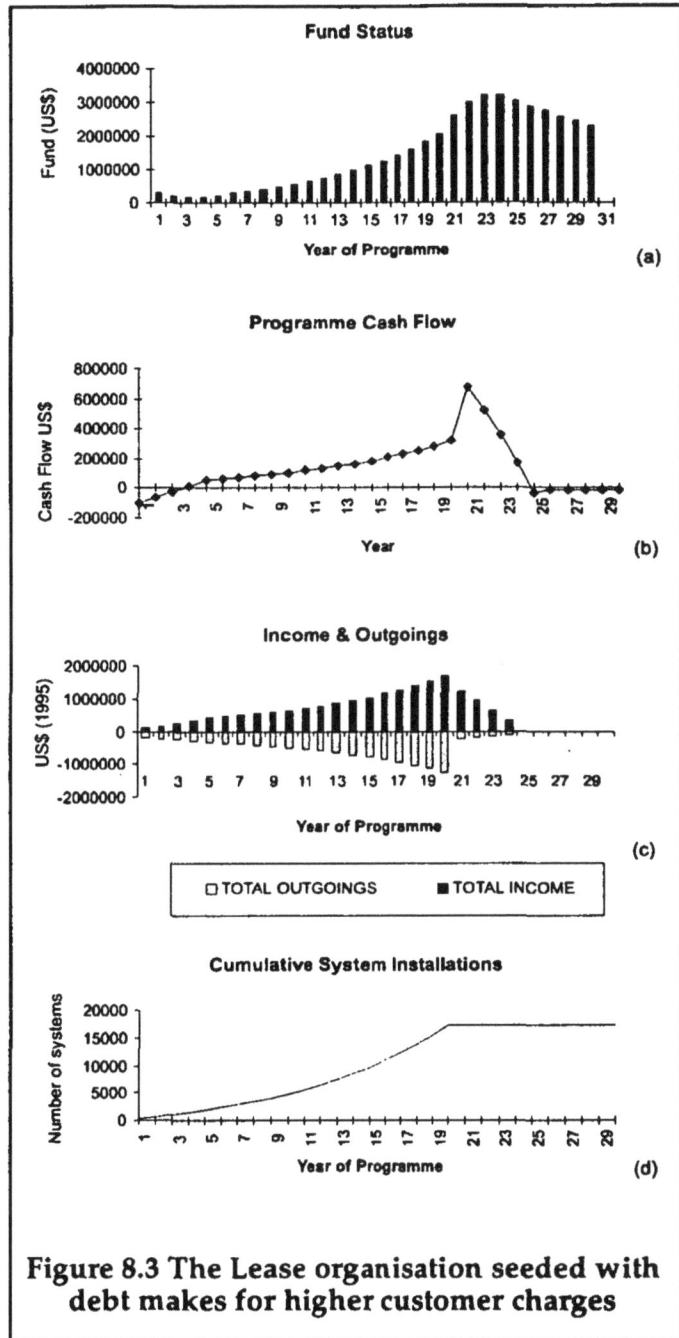

Figure 8.3 The Lease organisation seeded with debt makes for higher customer charges

Table 8.3 Parameters

Project Finance Summary	
Granted Seed Amount	$ 450,000.00
Debt Seed Amount	$ -
Grace period (yrs)	n/a
Debt repayment period (yrs)	n/a
Debt Interest Rate	n/a
Default Rate	2%
Customers' Finance Terms	
Principal Borrowed	$ 480.24
Repayment period (yrs)	5
Customer Interest Rate	30%
Customers' Annualised repayment	$ 197.18

with a comparable grant. In practice the fund manager should be able to recognise at quite an early stage that there is a problem and would take suitable action to remedy it.

Unless there is an additional subsidy to fall back on, and assuming there is no way of reducing costs elsewhere, the only course for avoiding such a decline is to account for these costs through higher customer repayments throughout the life of the programme. For instance, the problems demonstrated in this example could have been avoided by raising the interest rate charged to customers or by increasing the mark-up on the system cost.

If the same scheme had been operated with a customer interest rate of 30%, for example, the results would have been quite different as Figure 8.3 demonstrates. With the higher rate of interest, which corresponds to an increase in customer repayments of about $3 per month, the project profile appears to be much closer to that of the grant-seeded arrangement described in example 1. The project income comfortably outweighs the project expenditure, guaranteeing fund growth and ensuring that the full number of systems can be financed (once again based on the initial figure of 300 system sales in the first year of operation followed by 10% per annum growth).

Once the initial debt has been fully repaid (i.e. by year 8) the fund growth is substantial. Remember that the number of customers is identical to that in example 1, but each customer's repayments to the project funds have risen by almost $40 per year. The final fund balance at the end of the programme has increased accordingly.

Example 3 - Poor cost coverage

The effect of poor assessment of the costs of providing the credit service on a fund of this size is further illustrated in the next scenario. All assumptions and parameters remain identical to the base case described in example 1 with the excep-

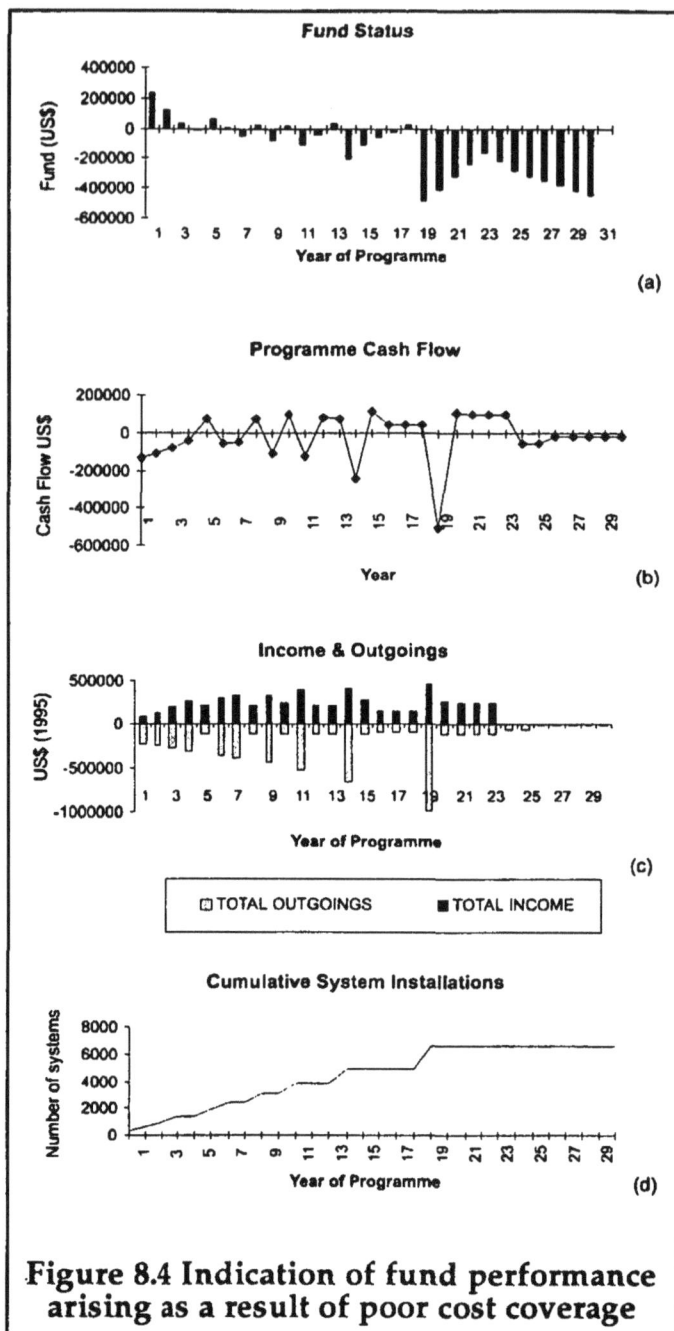

Figure 8.4 Indication of fund performance arising as a result of poor cost coverage

Table 8.4 Parameters

Project Finance Summary		
Granted Seed Amount	$	450,000.00
Debt Seed Amount	$	-
Grace period (yrs)		n/a
Debt repayment period (yrs)		n/a
Debt Interest Rate		n/a
Default Rate		2%
Customers' Finance Terms		
Principal Borrowed	$	480.24
Repayment period (yrs)		5
Customer Interest Rate		20%
Customers' Annualised repayment	$	160.58

tion that HQ salaries are doubled. This has the effect of increasing the fixed cost per borrower (i.e. reducing the profitability of each sale) and has particular significance when the number of customers making repayments at any time is small.

As Figure 8.4 demonstrates, there is again a marked difference between the performance of the programme in this instance and that of the base case scenario.

There is a considerable impact on the overall fund performance as the project experiences a continuous negative cashflow in the early years of the project which requires that some funds be transferred from the revolving wing to allow operations to continue. As example 2 highlighted, this reduces funds available for financing new systems which leads to a reduction in revenue in subsequent years. This is marked on several occasions by the fund actually slipping 'into the red'. The model will not allow systems to be purchased when the fund balance is negative. Only when revenue from customer payments for systems installed in previous years takes the fund back 'into the black' can more systems be financed. Taken to the extreme, if the outstanding revenue is not sufficient to return the fund to a positive balance, no more systems will be financed for the remainder of the programme. The overall effect then is to render the fund unsustainable.

To counter this problem (assuming the costs cannot be reduced) it is necessary to increase the profitability of the financed sale by raising the charge for providing the credit service or by increasing the mark-up on the system.

In this case there is some room for manoeuvre with the marginal system costs because the maintenance schedule is geared to meet replacement battery and lamp costs. The option exists to remove this benefit, but the wider implications to the programme of a decline in system performances or even system failure which might result from subsequent replacement with lower quality products should be carefully considered.

8.2 Conclusions

Because local requirements vary from one location to the next, it is difficult to offer a standard solution to meet the finance requirements of every potential RET customer in every area. Nevertheless, the key factor to remember is that the income to a programme - largely, though not necessarily entirely, from customer repayments - must equal or exceed the numerous costs of providing a credit service. Other points to consider are:

- harnessing low cost capital - either grants or low-interest loans - to 'seed' a revolving fund mechanism will help to limit the overall programme costs

- charging interest on loans to customers is the fund manager's key method for balancing the programme income with the programme costs

- as more customers are serviced, the cost per customer is theoretically reduced as many of the programme costs are independent of the number of clients. This effectively enables the finance provider to offer a cheaper credit service

- care should be taken to assess, and if necessary correct at the earliest opportunity, any potential bad debtors as these can be a drain on programme funds

- excellent staff and training are invaluable in limiting non- or late repayments

- contingency funds may be helpful to balance unexpected project costs

The next chapter looks at other parameters for implementing a successful financing scheme in developing countries.

Parameters for setting up a financing scheme　9

As can be seen in the preceding chapters, there are many issues which affect the implementation of a scheme for financing the purchase of small-scale renewable energy products at a local level. This chapter looks at a number of important issues and parameters in more depth, and analyses others which have not yet been discussed.

9.1 Determining the appropriate scheme

It is vital that the most appropriate type of scheme is selected. Before any analysis of technical and financial parameters is undertaken, the views of the potential users must first be established. If the community does not want a system, or households do not see the value of SHSs (for example), then there is little point in establishing a fund to finance such purchases, as it will have limited chances for success (Padmanabhan, 1988).

The most important consideration before attempting to initiate any financing scheme are the potential beneficiaries.

Willingness and ability to pay

It is equally important to ensure that the customers' wishes coincide with their financial capabilities. The main objective of RET financing schemes, as presented in this Guide, is to enable financially disadvantaged people to gain access to energy technologies which they would not normally be able to afford, that will offer life-improvements and possible income

generation capabilities. However, there is little benefit to the recipient in a scheme which tries to stretch too far. Trying to initiate a programme to finance sales of 80 Wp SHSs in a location where 30 Wp systems would be perfectly satisfactory (at a fraction of the price) may do more harm than good - it is likely that fewer customers would be able to afford the repayments or provide the deposit. Even if the larger systems are broadly affordable, the very fact that the financed sale price of the larger system is greater will mean that, for a given fund, fewer customers may be serviced.

It is important to remember that for the vast majority of the world's poor, any improvement in living standards, no matter how small, is better than none. It may be more appropriate therefore to think small but on a widespread basis, rather than assuming that bigger systems automatically mean better conditions.

Even in the poorest areas, there are some forms of saving (Otero and Rhyne, 1994; Foundation for Development Cooperation, 1992). In the absence of savings organisations, this takes the form of gold, precious stones, jewellery, livestock, land and other material goods. The lack of accessible facilities can also lead to the hoarding of money. These material goods and cash are then used in times of emergency or to pay for higher cost goods, such as PV systems. The Kenyan case study illustrates this point.

An assessment by GTZ of the demand and economic purchasing power shows that between 5 and 15% of rural households in developing countries can afford a SHS now, and would save by substituting their non-cooking energy requirements with a SHS. This represents the wealthier segment of the rural population, and translates, for instance, to 170,000 households in Morocco and 20,000 households in Namibia (Biermann et al, 1995). This is a significant immediate market, with a larger potential to follow as the market is established, infrastructures developed and more RETs financing schemes put in place.

Water pumping systems can be purchased through community financing schemes. Such schemes usually work on the 'user pays' principal

Photo: Siemens Solar

The savings propensity of an area may also be quite high: for instance in Indonesia, it is calculated to be between 10 and 20% of the household income (Otero and Rhyne, 1994; Thillairajah, 1994). However, without financial services tailored to the needs at a local level, local savings cannot be mobilised.

The debt repayment ability of the household, extended family or community needs to be assessed before a lending programme is initiated. Other-wise, a product and service may be. chosen which is out of the reach of the local community. Government concessions and fund requirements also need to be taken into account in assessing repayment ability.

A good starting point for calculating debt repayment ability is to find out what households currently pay for energy, such as candles, kerosene, batteries. Their income and its distribution over the year needs to be calculated, along with the probability of natural disasters in the area (such as crop failure). Full consultation with the community to ascertain their needs, expectations and financial requirements is also necessary before implementing the fund.

In Sri Lanka, the PV experience has shown that people are willing to pay in excess of their current energy expenditures for kerosene and batteries. This is because they value the quality of light, the cleanliness of the system and the ability to power small electronic systems. In one scheme over 2,500 systems have been installed on commercial terms, at interest rates of up to 34% per year and requiring substantial down payments. The repayments for these credit

There have been many successful dissemination projects for improved cookstoves in Asia and Africa.

Photo: D. Hall

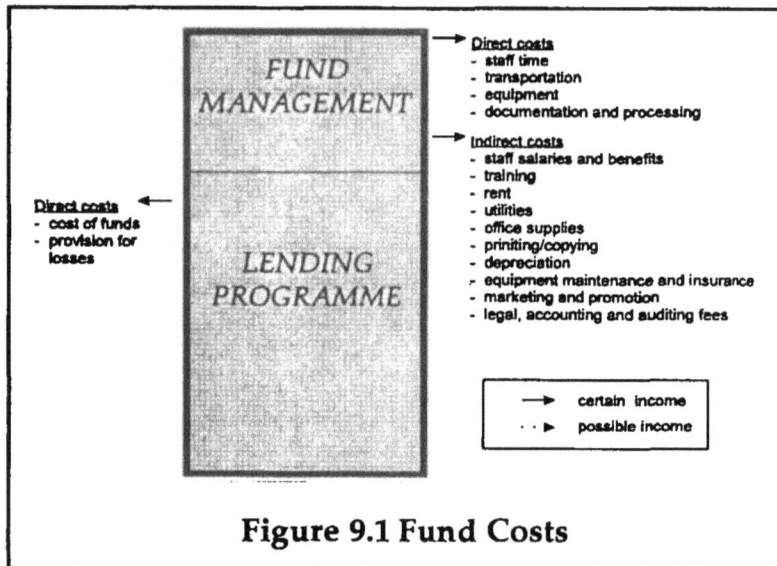

Figure 9.1 Fund Costs

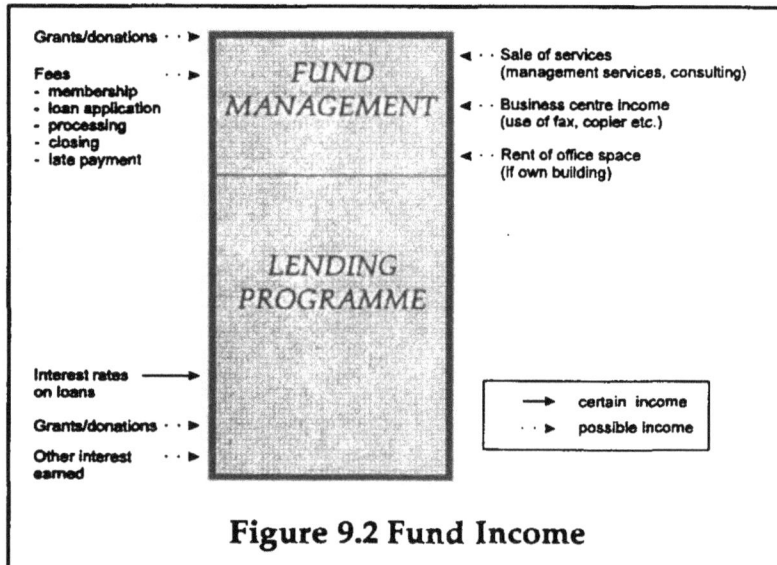

Figure 9.2 Fund Income

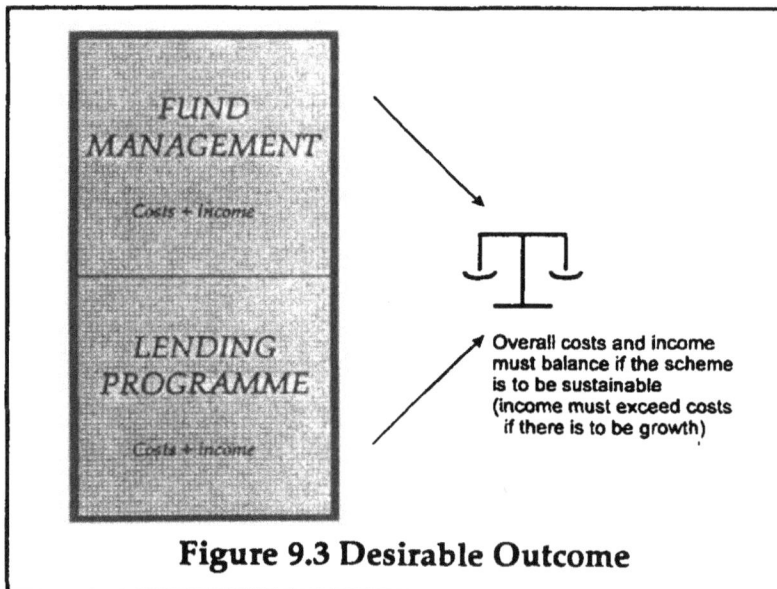

Figure 9.3 Desirable Outcome

schemes have been tailored to the disposable income of the customers, and the default rate is very low (Williams, 1991).

Assessment of appropriate technology

Hand-in-hand with the assessment of customer wants and needs should be an unbiased assessment of which technology is most suitable for satisfying client requirements. The performance of renewable energy technologies is obviously heavily dependent on the availability of a suitable energy source. If the assessment of customers' requirements indicates that better cooking facilities are needed, the potential alternatives for achieving this (i.e. solar cookers or improved biomass cookstoves) should be considered together. Are there abundant energy resources for both technologies? If not, the viability of any programme aimed at selling such technologies should be questioned.

Drawing on previous local experience

Once the users' needs, and the potential for RETs to fulfil them, have been established, it is worthwhile evaluating the current structure of the local community. There may already be some form of financing scheme operating, be it formal or otherwise, which could be adapted for the purpose of financing RETs. There is generally little to be gained from 'reinventing the wheel'.

It is also essential to consider the needs of the financiers. Only then can a decision as to the most appropriate mechanism (revolving fund, ESCO, lease organisation etc.) be made.

9.2 Implementation and infrastructure development

Costs and Income

The scheme will have two types of costs: direct and indirect. Direct costs incurred by the fund include salaries, collection of repayments, equipment, documentation, loan processing, disbursement costs, borrowing capital and provision for losses etc. (see Figure 9.1). The size of direct costs will vary according to the size of the fund. Indirect costs also vary according to the size of the fund, but are more variable, and can be increased or decreased more easily.

Income to the fund is also variable. Income to the lending programme includes interest on loans, plus any additional income such as grants/donations and interest earned on other deposits. The fund management can also derive income from loan administration (fees), and has the ability to earn income from the provision of other services, such as consulting services to businesses, renting out additional office space, providing business centre services etc. (Figure 9.2).

However, the fund must be self-sustaining over the medium to long term (Figure 9.3). Costs must be less (or not greater than) income.

Financial relationships

The relationship of the fund with other service providers is important. These include:

● formal financial organisations
● similar financial organisations
● (potential) lenders to the fund
● technical service organisations

As shown in chapters 5 and 6, there are a variety of different models for interdependency with other financial institutions. The appropriateness of the fund will depend on issues such as the financial regulations of the country, experiences to date with this type of lending mechanism, financial sophistication of the new organisation and local conditions. If there is no savings mechanism within the new fund, then it may be necessary to link it with a local or regional banking network. If the scheme includes savings facilities, then the link may be a formal one facilitating access to loan funds. Figure 9.4 shows the possible flow of commercial loan funds from international sources (such as the World Bank, benevolent foundations and so on) to a formal banking institution at the national level, and disbursement down to a local level.

There are a number of other possible scenarios for loan disbursement:

● local savings may be sufficient, if linked to an existing savings fund (either formal or informal)

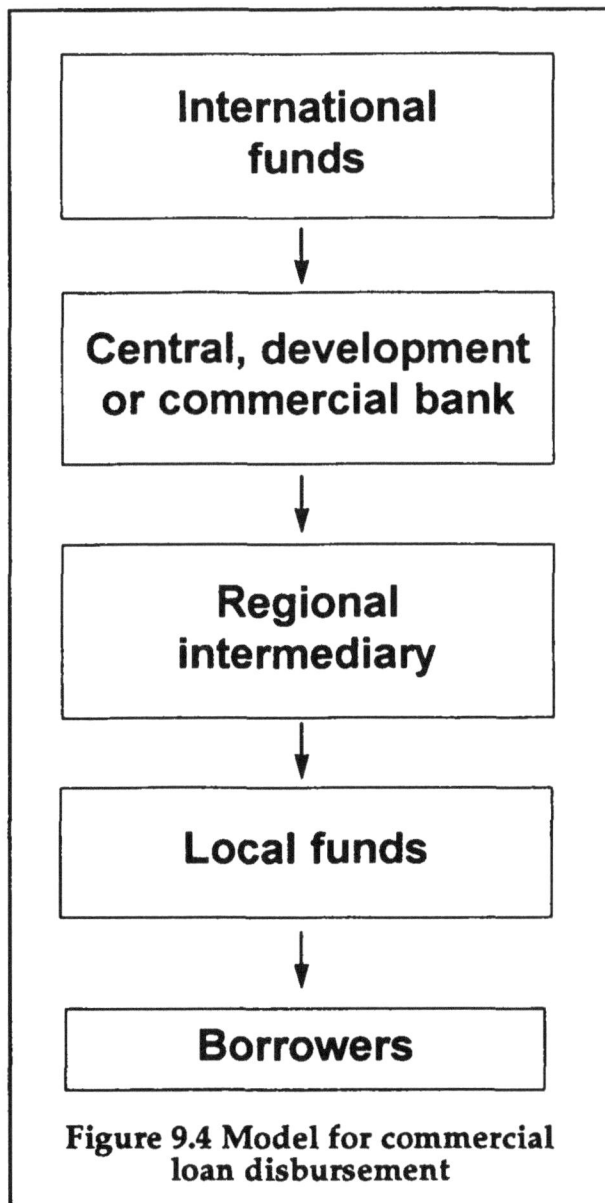

International funds

↓

Central, development or commercial bank

↓

Regional intermediary

↓

Local funds

↓

Borrowers

Figure 9.4 Model for commercial loan disbursement

- international sources of funds may prefer to loan directly to a local level intermediary. For instance, the Rockefeller Foundation makes loans and grants to both SELF and Enersol Associates. Much of this is injected directly into local level lending NGOs

- formal banks may lend directly to large numbers of small borrowers

- national banking organisations may disperse the loan to a number of regional intermediaries, who in turn may on-loan to a number of local organisations

Before making initial approaches to formal banking institutions, any new scheme should aim to have its objectives clearly defined, the business plan developed, the short-run and long-term calculations for the sustainable management of the fund prepared, and a policy of self-help, not altruism, declared. If these documents and polices are well developed, it will increase the likelihood of joint participation by the formal institutions.

As the scheme develops, it may become desirable to change affiliations, or to establish new ones. In some cases, networks of financial organisations have been set up (for instance, Women's World Banking and Accion International), which lend support through knowledge dissemination between affiliated organisations. For a new organisation, this means access to existing structures and procedures which can be adapted to local conditions. This reduces risks of putting in inappropriate structures and methodologies or having to pay the cost for 'reinventing the wheel'.

Regardless of the type of structure which is set up, tight financial and administrative control must be maintained by the scheme.

Portfolio Management

The ideal investment portfolio for any fund is a diverse one: different sizes of loans, different duration of loans, and loans for a variety of uses (e.g. seeds, capital equipment, consumption).

Looking particularly at a PV fund, it is important to assess the following issues:

- is it preferable to have a range of PV applications within the loan portfolio (e.g. SHS, water pumps, small enterprise lenders), or can the fund be sustained by lending for one type of system (e.g. SHS)?

- should there be a diversity of borrowers (e.g. from the local community but not all reliant on crops for their income)?

- should there be a diversity of loan size (catering for the wealthier as well as poorer sections of the community)?

Each loans officer will be responsible for a number of loans. This includes screening and assessing the credit-worthiness of potential borrowers; establishing the loan and the method and rates of repayment; undertaking required internal administration; and otherwise making sure that the loans within his/her responsibility are working in a profitable way for the fund.

A lending portfolio may include lending for a number of uses, such as stock watering system
Photo: Siemens Solar

When a borrower takes out a loan, the loans officer will

A well installed and maintained system reduces the potential for dissatisfaction with the system and thus lowers the likelihood of loan default. Photo: Sollatek

meeting local needs, the ability of loans officers in screening clients and their ability in portfolio management, and the scheme's overall standing in the community. It is important that each loans officer is aware of the performance of borrowers within the portfolio, and maintains a strict control over delinquency and defaulting borrowers. If these two areas are allowed to slip, then the fund will become unsustainable and fail. Figure 9.5 summarises the key factors that can help to ensure low customer default rates.

have evaluated his / her ability to repay, and will have taken security against the loan. The borrower must be aware of sanctions which will be taken if he / she defaults. If a defaulter does not pay, then this mechanism should be instituted immediately, as otherwise other borrowers will hear about this laxity and the default rate is likely to rise. Schemes should have access to legal expertise, such as lawyers who can administer loan collection once it becomes necessary.

Low default rates are a good performance indicator of the scheme's achievement in

Another important element of portfolio management is making sure that the objectives of the programme are met, such as lending to a certain economic segment of the population. This is particularly important with relation to disbursement of loans to the poorer segments of the community, who typically have had little or no interaction with formal or semi-formal borrowing and may be illiterate and thus unable to read loan application forms. If this section of the population is important for the fund, then every effort has to be made to overcome the barriers posed by low literacy skills. Actions taken by the fund can include more time spent with individual customers, talking

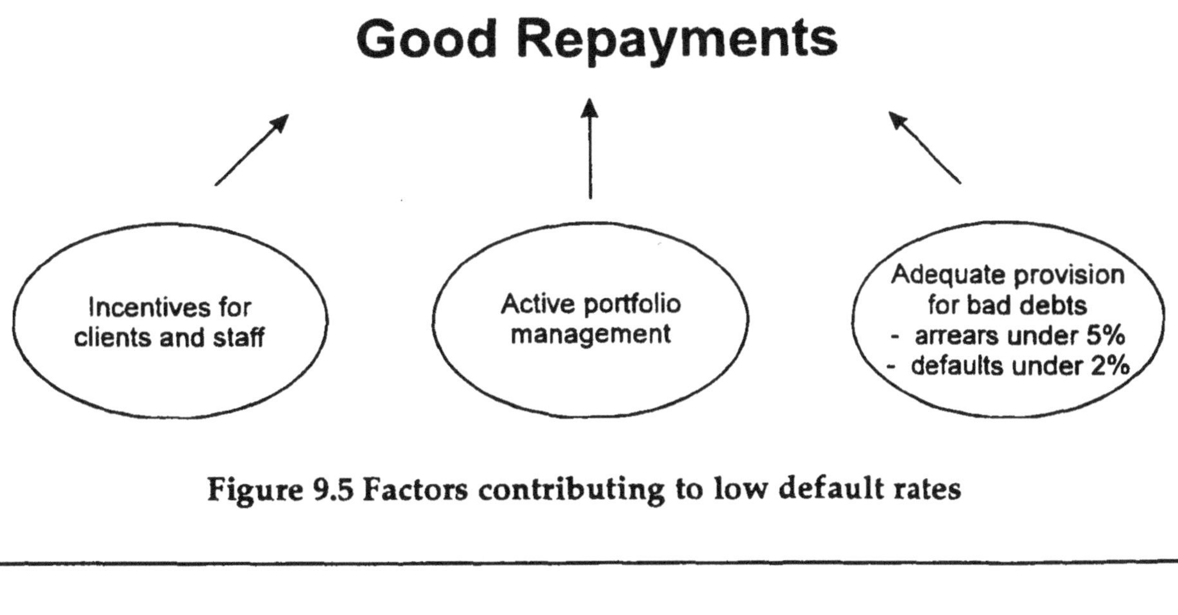

Good Repayments

Incentives for clients and staff

Active portfolio management

Adequate provision for bad debts
- arrears under 5%
- defaults under 2%

Figure 9.5 Factors contributing to low default rates

to them about the responsibilities of borrowing from the fund and assisting them to fill out the necessary forms.

Administration

Any financial organisation needs good management. A process of review and evaluation of lending procedures, staff training, defaults, delinquency, size of fund etc. must be undertaken regularly, otherwise problems being experienced by loan officers may pass unnoticed until too late and / or the fund may lose its usefulness to the community. In general, once a fund has been established in an area,

Installing a SHS in Kenya   Photo: Mark Hankins

potential borrowers will gradually become more knowledgeable (Padmanabhan, 1988; WWB, 1993). Thus, what begins as a small revolving fund for SHSs, with no savings facilities, may find that the demand grows to include larger systems and / or savings facilities. If this change in demand is not met, then the fund will no longer service the financial needs of those it was set up to serve.

Internal financial checks and balances have to be established when the fund is set up, and those procedures closely followed in order to ascertain the operational performance of the

fund. If administration costs become too large compared to the income to the programme, then these costs must be reviewed and excess expenditures curtailed. Some of the key issues that can assist in providing a cost effective lending scheme are summarised in Figure 9.6.

Service charges of the fund must take into account all costs, including interest rates, inflation, administration, profits/dividends, write-offs for bad debt etc. If this is not done, then the self-sufficiency of the fund will be compromised.

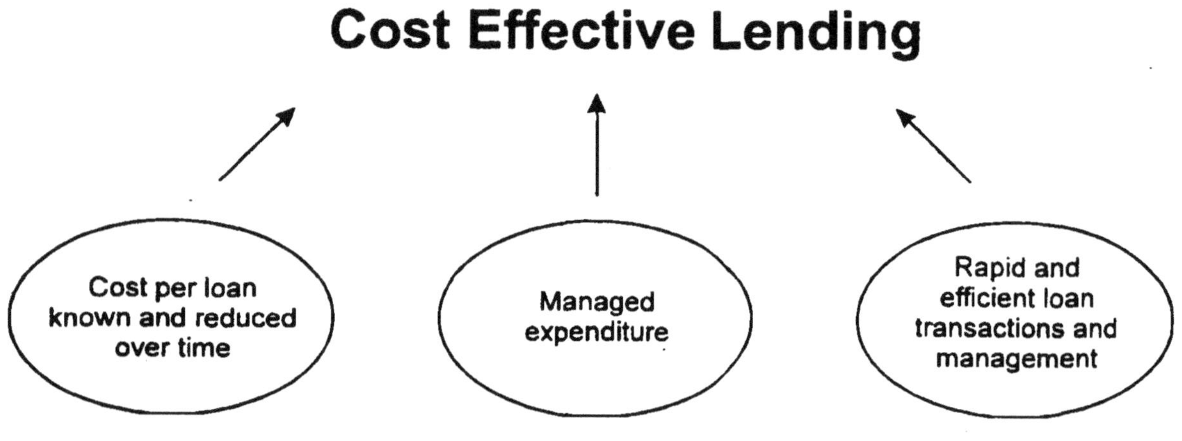

Cost Effective Lending

Cost per loan known and reduced over time

Managed expenditure

Rapid and efficient loan transactions and management

Figure 9.6 Issues for cost effective lending

Lending risk

The lending of money always has an associated risk of default. With most lending organisations, this risk is amortised into the interest rates charged to customers. One way to lower the risk of default is to require collateral of a similar or higher value to the loan and / or to require compulsory savings. Every effort needs to be made to minimalise the financial risks involved for organisations lending to the fund, such as spreading the risk across a number of financial organisations and ensuring that tight financial management controls are established and adhered to within the fund itself (Thillairajah, 1994; Germidis et al, 1991).

Default rates on loans are least when there is peer lending, i.e. the people who are borrowing from the fund are from the one community. If the fund also provides for local savings, it becomes an even greater social responsibility to repay. There are successful examples of PV lending which have not been adversely affected by risk. These include schemes facilitated by Enersol Associates in the Dominican Republic and Honduras; schemes set up or facilitated by the Solar Electric Light Fund in Sri Lanka, China and India; and the electrification of many parts of Indonesia (see case study).

Risk abatement is also important when assessing the possibilities for finance for a project. In the case of larger projects which include RETs financing schemes (such as the GEF PV project in India), participation by the World Bank can be critical in mitigating both political and credit risks (Hass and Bender, 1996).

Technical services

RET financing schemes have a number of choices in the provision of technical services. If the organisation is an Energy Service Company, it will supply all the services, including all aspects related to the supply of equipment as well as finance. If the organisation is an NGO which prefers to deliver financial services for RETs financing only, then it will require the private sector to develop the local technical infrastructure. This may be in direct co-operation with the fund (for instance, installers contracted to the fund) or in parallel (i.e. no co-operation but private sector developing to provide technical support).

Getting the technical detail right is just as important as getting it right financially. If the wrong equipment is specified, or installations and maintenance are not undertaken in the correct manner, systems will not work properly and the users (the borrower) will become disillusioned with the system and be likely to default on payments. If equipment is returned to the fund, then costs are also incurred through lost revenues and incorrect expenditure on equipment.

Whatever the technical arrangement offered, a close watch must be kept on the technical services, and management systems put in place to ensure quality products, installations and after-sales service.

Technical questions which need to be addressed include:

- all installers and maintenance personnel must be expertly trained

- at least one person within each local community should be expertly trained to undertake routine maintenance and repair

- quality products with guarantees must be used within the programme

- users need to receive adequate information on the parameters of their system before purchase, and receive training on how to use the systems and undertake basic maintenance (such as topping up batteries) at the time of installation

- an adequate equipment supply infrastructure needs to be established and maintained

The cost to the programme and the quality of this service must be monitored and evaluated at routine intervals so that technical disasters and cost over-runs do not occur.

Potential users need to be educated about system performance capabilities before they apply to borrow funds. If the system does not

Training engineers, Guyana   Photo: IT Power

perform up to expectations, disillusionment will result.

Questions which potential users need to ask before agreeing to purchase a SHS include:

- what does the system power? (e.g. TV, lights, cooker etc.)

- how many of these can it power?

- how long does it power the equipment for?

- what happens if the system is highly used or not used much at all?

- what type of equipment is needed? (e.g. deep cell battery, 50W modules, four energy efficient lights, regulator, wiring)

- are there any additional costs? (replacement of batteries, maintenance)

- who has to pay for any additional costs? (fund provider, private company affiliated to the fund or the user)

Even if the fund is not responsible for the disbursement of equipment, it should still have information sheets with questions and answers about the RETs financed, for the borrower to read before he/she signs up for the loan. This can assist in reducing the risk of disillusionment during the period of the loan.

The technical infrastructure which needs to develop to support the widespread dissemination of RETs equipment includes suppliers, trained installers and maintenance repair technicians, and, as the programme grows, local manufacturers of components and equipment. The technical infrastructure needs to have an installation density high enough to support continuous local supply of equipment and provide income for at least one local trained technician. As the market grows, this can expand to support more employment opportunities, such as equipment assembly and manufacture and more maintenance and installation requirements. This increases the economic wealth of the local community, and brings new skills into the community.

9.3 Training

Sound training of financial staff is critical to the success of a fund. Training costs can be as high as 24% of total costs, though tends to average around 5% (Otero and Rhyne, 1994). Obviously training costs will be highest during commencement of operations and during phases of expansion. Due to the high investment in well-trained staff, it is important that staff are selected carefully at the commencement of the project, retained for as long as possible (high staff turnover will increase training costs), and given training upgrades as required.

Training of technicians is also critical - if installation and maintenance of systems are part of the scheme, local technicians must undergo full training accreditation, otherwise incorrect procedures will be followed and systems will fail.

9.4 Monitoring and evaluation

The ability to maintain strict financial control must be factored into the structure of any

92

fund. This refers not only to portfolio management but also to general administration and any activities taken outside the lending programme, such as technical installations and additional services.

There is international debate about the criteria for evaluation of the success of a financial programme (Hilhorst and Oppenoorth, 1992). Evaluation criteria need to be developed in order to assess the scheme's performance. Criteria could include:

- the percentage of the target population which is reached
- types of loans which are dispersed
- size of loans
- analysis of the rate of on-lending
- size and frequency of deposits (if relevant)
- loan collection and administration costs

Whatever the format of the financing scheme which is established, the method of evaluation must be part of the fund's operations. This will require the ability to change as the fund evolves, otherwise the original performance indicators may give an incorrect evaluation of the scheme.

9.5 Effect of market enablement actions

As discussed, the long-term financial sustainability of a fund is paramount. Facilitating economic development and higher standards of living in the rural areas needs as many households as possible being able to access suitable credit and savings facilities. Therefore, market enablement actions need to be assessed carefully with regard to their short and long-term effects. If subsidies and grants are to be applied to the fund, these must be administered in such a way that the fund will not become reliant on them in the long-term. They should be used to develop infrastructures and invest in human resources, rather than being used to subsidise interest rates. If grants or cheap loans are used to capitalise the fund, then the interest rate to borrowers should still reflect present and future costs of the fund. Otherwise, they are providing cheap capital for a small minority of the people in the short term.

There are four levels of financial sustainability:

i. Grant or soft loan

A grant or soft loan can be used to seed a revolving fund or credit co-operative. These are currently provided for PV financing to the Tuvalu Solar Electric Co-operative Society, Enersol Associates for activities in the Dominican Republic and Honduras, and to the Solar Electric Light Fund (India, China, Sri Lanka etc.). If a fund is totally reliant on these mechanisms, then it will not find self-reliance.

ii. Below-commercial rates of interest

Development banks and international sources of soft loans can assist the fund in providing lending at lower interest rates. For instance, agricultural development banks often have access to subsidised state loans. Another example is the PV revolving fund, administered by IREDA, India, as part of its Global Environment Facility renewable energy programme. This is a useful market mechanism to 'kick-start' a programme or project. However, it should have a clearly-defined timetable, and the borrowers should know that it is a special rate for them, not a debt which will not have to be repaid. If the subsidised rate is not transparent, then problems such as high default rates and delinquency could occur.

iii. Most subsidies eliminated but dependency in one or more areas

This is the most usual scenario for provision of funds to lower income sections of the population. Here most of the subsidies have been eliminated, but it has lower than commercial interest loans and lower public charges.

For instance, the Grameen Bank of Bangladesh has a low cost of capital, and the Badan Kredit Kecamatan of Indonesia uses grant support to pay for administration costs.

iv. Totally self-financed and growing

This is the optimum state for a self-reliant fund. Total self-reliance occurs when the fund has adequate capital, it can repay its scheduled debts, and has enough surplus to expand the fund. This is how commercial financial operations work around the world. There are examples of this in development areas: for instance, within the Women's World Banking operations (WWB, 1993). Whilst there are no such operations supporting renewable energy at this point in time, it is the target to be achieved in the longer term.

9.6 Issues for subsidised credit

There has been much debate across development sectors about the merits or otherwise of subsidised credit. Prior to the mid-1980s, subsidised credit loans were a normal part of many aid and development programmes. In the 1980s the use of subsidies was challenged and many studies were undertaken to ascertain the real effects of subsidies on development programmes. Today, the prominent view is that long-term subsidies for sustaining a fund are not acceptable, as they have negative effects (Castello et al, 1991; Otero and Rhyne, 1994; Thillairajah, 1994). These effects include the following:

- low-priced credit can generate excess demand, necessitating rationing. This can lead to corruption and political intervention, and reduces the borrowers repayment level. It also usually means that those who need it most (i.e. the illiterate poor) are denied access

- low interest rates mean low operating margins for the funding organisation, and reduce the services which they are able to offer. If too low, they can reduce financial sustainability

- default rates are higher where there is no community involvement (i.e. peer group lending) in the fund

- subsidies can lead to over-optimistic price expectations, and are very hard to phase out as purchasers want prices to remain low

- subsidised credit can also lead to avoided repayment, either through borrower default or cancellation of the debt (e.g. in the run-up to an election, the debt may be

cancelled by the government in order to 'buy' votes)

- subsidised credit perpetuates paternalism (i.e. handouts and dependency), unless it is transparent and has a timetable for discontinuation

- low interest rates can indicate to borrowers that a lending organisation is 'soft' on repayments, with the possibility that debts will be 'forgiven' in the longer term. This leads to avoided repayments and a high default rate

Money is lent at exorbitant interest rates, particularly in the informal sector of developing countries. This typically ranges from 10% a month to 20% a day. Commercial lending rates are already cheap by comparison, so, in real terms, subsidies may not be necessary.

Direct subsidies were given for PV in the aid programmes of the 1970s and 1980s, when many 'free' PV systems were given to beneficiaries. However, these programmes had many negative effects (Biermann, et al, 1995; Hoffman and McNelis, 1986; IT Power, 1996c):

- beneficiaries often were not consulted as to their needs and energy requirements, and thus had no sense of ownership of the systems

- beneficiaries often did not have the financial resources to maintain the equipment donated

- an unrealistic price expectation for PV was created. PV companies/traders found it very difficult to sell to rural populations in developing countries, as the people's expectation was that they would receive it free from an international aid agency or NGO

- as systems were given away to the local communities and households, they had no real value; the recipients did not have to pay for them in any way and therefore did not value them

A better situation is to combine normal financing mechanisms with grants for infrastructure development, rather than heavily subsidising interest rates. This means that seed capital and/or equipment can be used to start the scheme, while training and dissemination activities can be paid for by outside agencies, and commercial interest rates (or just below) applied.

Financial organisations need to charge commercial, or near commercial, rates of interest to borrowers to make sure that they cover their own lending and administration costs. If the lending institution cannot cover its own costs, then it makes the programme/project unsustainable and reliant on external financing, which is undesirable for sustainable long term development or bringing self-determination to local communities. Even if the government, for instance, makes a commitment to bring PV electricity and economic development to rural areas and heavily subsidises the programme, it does not necessarily follow that these systems should be 'free'. Instead, community commitment, such as that required by the PRONASOL programme in Mexico (Huacuz et al, 1994), is a way of ensuring that the systems have a value for the recipients, even if the payment is 'in-kind', such as installation or maintenance work.

9.7 Conclusions

There are many issues which will affect the establishment of a financing scheme. These vary in impact, depending on local conditions and the type of organisational structure chosen for the programme and/or fund. A check list of considerations for establishing a programme is shown in chapter 10.

The financial self-sustainability of the fund is critical. Increasing the services which it can provide and expanding its loan disbursement capabilities are also important issues.

On the technical side, there are many actions which can be taken to ensure that quality

Water pumping systems mean potable water for multiple use. This means that the water fetchers (usually women and children) can do other activities (e.g. paid work, attend school etc.)   Photo: IT Power

products are used, that installations are by trained technicians, and that there is high consumer satisfaction with both the systems and the technical support.

The credit programme must meet the long-term needs of its local clients in a way which both provides the clients with appropriate services and ensures the profitability and continuity of the fund.

There are a number of key factors for successful financial implementation:

- there must be strong local support for the financing scheme

- the programme and fund must have clearly defined policy and objectives, based upon self-help at a local level

- there must be a well-thought-out business action plan designed to achieve these

- there needs to be clear lines of responsibility and reciprocation for all organisations participating in the programme

- staff must be well-trained in loan disbursement and administration

- within the fund, there must be clear guidelines for lending policy, loan disbursement, repayment collections, minimisation of defaults and portfolio management

- reporting procedures must be set at the outset of the fund and developed as the fund grows

- exposure to risk must be minimal

- the programme must have a strong knowledge of the financial needs of the local potential borrowers, and the economic and social issues which will affect their capacity to borrow and make repayments

- collaboration with similar organisations can bring many benefits

- interest payments must be enough to cover costs of the programme (including general administration and loan repayments as well as loan administration)

- if there is a technical component, it must either be self-supporting, or the cost of it should be factored into the interest rates and administration charges

- if the programme does receive subsidies or grants, then a long-term lending strategy must be in place for these to be phased out

- soft loans or subsidies must not prohibit the development of other financial lending schemes in the sector

- the portfolio ideally should include both large and small loans, and borrowers with a diverse income background (the second requirement is the most important to hedge against natural disasters in an area, such as crop failures or drought)

- adequate provision needs to be made for inflation, otherwise the value of the fund will diminish

- a good relationship must be maintained between the programme (and, in particular, the loan officers) and the community to which it is providing services. The loan officers need to know the status of the local economy, and the issues impacting upon it (both short and long term)

- appropriate collateral arrangements should be arranged, which do not prevent lower income people from borrowing from the fund. At the same time, these arrangements must also protect the lending organisation from 100% risk exposure

The programme must also be flexible enough to change if necessary. For instance, moving from informal lending to the semi-formal or formal sector, with savings facilities incorporated into the programme.

Project checklist 10

There are many different issues which need to be answered and resolved before any scheme for financing renewable energy projects in developing countries is established. This chapter takes the main issues, and summarises them into a checklist format. The list is by no means exhaustive and indeed it is expected that other, local questions should be listed. It is also expected that some of the answers to the questions posed will not fit into the space provided. When this is the case, a more comprehensive investigation and report should be required due to the importance of the questions. An example of this could be: What have been the experiences with rural credit, money lenders etc. in the area?

The aim of this chapter is thus to act as a reference for the types of questions and issues which are important for setting up a scheme. Information needed to understand these questions is contained within the preceding chapters, and in the further reading listed in the Bibliography.

IDENTIFYING THE NEEDS AND FINANCIAL RESOURCES OF THE BENEFICIARIES	COMMENT
What are the energy services the beneficiaries require? (e.g. lighting, TV, health care)	
How much can the beneficiaries afford to pay each week or month?	
Are there times that they can afford to pay more (e.g. after harvest)?	
What have been the experiences with rural credit, money lenders etc. in the area, and how can the experience be used?	
IDENTIFYING APPROPRIATE RENEWABLE ENERGY TECHNOLOGIES	COMMENT
What has been the experience with renewable energy technologies in the locality to date?	

Do commercial distributors of renewable energy equipment service the area?	
Which technology is the most economic in the area (on a life cycle cost basis)?	
Which technology or system is financially most viable for the beneficiary?	
Can adequate servicing and spare parts supply be established?	
Has there been international experience with the chosen technology? If not, is it wise to use it?	
Are the systems good quality and do they have proper guarantees?	
IDENTIFYING THE MOST SUITABLE TECHNOLOGY DISSEMINATION MODE	**COMMENT**
Are there local NGOs or co-operatives experienced with renewable energy technologies?	
Would the electricity utility be a viable dissemination vehicle?	
Do Energy Service Companies (ESCOs) exist?	
If not, would ESCOs be viable in the area?	
Are there existing supply lines in place that could be used? (e.g. livestock or agricultural companies)	
What mechanisms are the existing renewable energy distributors using for dissemination?	
Is there an existing local RETs infrastructure? If so, which technologies does it service and does it need strengthening?	
Is it advisable to employ local consultants to advise on a dissemination plan and other activities?	
ESTABLISHING FINANCIAL DISBURSEMENT SCHEME	**COMMENT**
Review existing set-up of rural credit, savings and rural banking	
What are the costs and overheads associated with these?	
What intermediaries are working with banks and borrowers?	

Are there plans to establish a National Renewable Energy Fund?	
Is international seed money or grant aid available for associated activities (e.g. training or standard development)?	
What are the requirements of donors for any grant or seed money?	
Is there concessional financing available? (e.g. GEF)	
Will intermediaries be needed? (e.g. community based organisations for payment collection services)	
What is the new fund's lending policy?	
ESTABLISHING FINANCING PARAMETERS	COMMENT
What will the transaction costs be?	
Over what period will repayments be made? How does this compare with the life of the equipment?	
Determine payment levels after thorough cash flow and sensitivity analyses	
Establish default policy. Who is guaranteeing repayments?	
Establish incentive schemes	
DEVELOPING THE BUSINESS PLAN	COMMENT
Who is the owner/principal stakeholder?	
What will be the cost of the management infrastructure? (revenue collection, training, specification and standards development, quality control, staff overheads)	
What risk exposure does each stakeholder have?	
What will be the Internal Rate of Return?	
Undertake sensitivity analyses on future cash flows (in respect of defaults on income and costs of maintenance, revenue collection, staff, future purchases etc.)	
What size of scheme is most appropriate?	
Is the scheme permanent or transitional?	
Will the beneficiaries lease or purchase the equipment (or both)?	

FINANCING RENEWABLE ENERGY PROJECTS

Where will set-up funds come from and under what terms?	
What will be the effects of inflation? How will these be countered?	
How will borrowing risk be minimised?	
How will the lending risk be minimised?	
How will exchange rate risks be minimised?	
ESTABLISHING THE MANAGEMENT INFRASTRUCTURE	**COMMENT**
Select financially experienced staff and train as required	
Establish networks with other organisations	
Design revenue collection	
Establish specifications, tendering policy, equipment sourcing policy, quality control etc. as necessary	.
Develop marketing and promotion plans	
Establish management responsibilities	
Establish equipment maintenance programme	
IMPLEMENTATION	**COMMENT**
Is the portfolio diversified enough?	
Are the lending parameters working?	
What is the lending rate per month? Is it high enough?	
What is the default rate per month? Is it too high?	
Are interest rates high enough to cover costs?	
Are there costs which were not budgeted for?	
What can be improved or needs modification?	
MONITORING, EVALUATION AND ADJUSTMENT	**COMMENT**
Establish project monitoring systems	
Establish evaluation criteria	
Adjust financial and implementation parameters as necessary	

PV HOME SYSTEMS IN KENYA

Between 1985 and 1996, Kenya has nurtured a healthy solar PV market, with between $2 and $4 million worth of equipment being sold each year. At least half of this market consists of home lighting systems (SHS) between 12 and 200 Wp in size. This portion of the market is driven by strong rural demand, and is supplied by a few major distributors operating through scores of outlets, many of which are located in provincial centres in rural cash-crop areas. There are now more PV systems installed in rural Kenya (over 40,000) than there are connections under the Kenya Power and Lighting Company's Rural Electrification Programme.

There is also a strong market for systems in remote, off-grid areas such as the semi-arid north and pastoralist areas. Systems in these areas are mainly financed by donors, missionaries and government agencies. The larger systems in this market segment tend to be supplied by Nairobi-based companies that, unlike the less-sophisticated rural agents and technicians, can design, bid on and supply for contract tenders for international customers.

The growth of the SHS market is of particular interest, because it occurred in the absence of any 'project' effort and was driven almost completely on a local basis.

Starting in about 1984, the commercial PV market grew based on cash sales of systems to rural customers. In several areas, rural revenues are (by African standards) high, due to revenue from cash crop sales from coffee, tea, maize, quat (a mild stimulant popular in Arab countries) and other agricultural commodities. Demand for lights, radio power and televisions - in the face of slow expansion (and politicisation) of the Kenya Power and Lighting Company's Rural Electrification Programme - caused many farmers, teachers and business people to seek alternatives to grid connection. Until the mid-1980s, the only alternatives to the grid were small generator sets and centrally recharged automotive lead-acid batteries.

Between 1984 and 1987, several small-scale pioneering demonstration and training efforts in such areas as Meru, Kisii and Maragoli stimulated interest in the technology among potential customers. During these catalysing efforts, local technicians were trained to install systems for higher-income customers (i.e. head-masters, business people, city-based employees with rural homes). Although schools provided excellent demonstration venues, rural homes turned out to be the largest initial market. Local electricians and merchants were quick to see that PV could take up much of the niche which, at that time, was dominated by generators. The cost of a PV system, between $500 and $1000, was often less than the initial cost of a generator. Demand for PV increased rapidly, and the number of solar distributors in Nairobi grew from three in 1984 to over 15 by 1995.

Initially, all PV transactions in Kenya were cash sales. Factors that fed the commercial growth include the following:

- Strong cash economy: The initial market tended to be rural people who had built permanent, comfortable houses, and who were able to invest in solar power. Customers were often leaders and innovators with respect within the community. Systems were built in clusters as the first installation would often 'sell' several others in the immediate areas.

- Rural presence of installers and retailers. The presence of installers who could guarantee performance and after-sales service sustained user credibility in the technology. Even if customers bought equipment in Nairobi to save money, they often used local electricians to maintain and service systems.

- Availability of equipment. Unlike neighbouring countries Uganda and Tanzania, in Kenya (Nairobi) there was a constant stock of PV modules, batteries, regulators, low-voltage fluorescent lamps and BOS equipment. Equipment was stocked by small up-country dealerships, across-the-counter outlets in towns and major PV distributors in cities. Locally manufactured lamps, solar batteries and charge regulators reduced system prices and contributed to the healthy industry.

- High demand for TV and radio. Rural people wanted to have the same services as city people and TV is both an educational tool and modern 'status' symbol. Twelve volt DC Chinese TV sets were available for about $50, and the television network (again, unlike Tanzania and Uganda) reached most of the populated areas of the country. Tens of thousands of people already carried batteries back and forth between rural homes and mains charging centres so that they could run TVs and radios. Fixing a PV module to the battery to reduce or eliminate the need for distant charging was an attractive improvement to TV power systems.

Finance efforts

In general, Kenyan rural finance projects have, until recently, had a poor history and consequently there were few early efforts to finance PV home systems. Commercial PV companies work on a cash basis and do not want to lock up their capital in risky, widely scattered loans. Banks still view PV as a new technology and are wary of investing in PV for rural customers. However, as the 'cream' of the market (i.e. that portion that can pay the cost of buying and installing a full home system) diminishes, companies and retailers are seeking ways to decrease up-front costs of solar, one of the most difficult hurdles in expanding the SHS market. This will allow expansion into the significantly larger middle-income market.

As of 1996, the most common sales transaction was payment on installation. This involves partial payment of the agreed-upon purchase price upon signing of a contract, with the remaining amount paid upon completion of installation.

This allows the dealer to buy components and assures the customer that the installer will not 'disappear'. In many cases, the customer buys the modules, batteries and major BOS equipment in the city, and allows the rural installer to purchase minor equipment - this cuts down systems costs.

Informal financing is fairly widespread among provincial dealers who know and trust their customers. One Meru-based dealer has, since 1989, offered financing to trusted customers who have regular incomes (i.e. monthly payslips). Customers pay 50% down, and then pay 25% upon installation of the system. They then have three months to pay the remaining 25% without interest. If they take more than three months, they pay 15% interest on the loan, and after six months of non-payment, the dealer comes to repossess the module. To get it back from the dealer, the customer must pay the balance, interest and a reconnection fee. This has proved a successful marketing tool, as this dealer (who sells from four to ten systems per month) provides the loan service to more than a quarter of his customers. In the case of his business, no customer has defaulted.

Hire-purchase schemes have been tried in the past and are again being re-introduced. In the late 1980s Argos, a hire-purchase firm selling consumer amenities, introduced a 20 Wp amorphous solar package which sold in the hundreds. The scheme ended because the Argos company itself closed, but many of these systems failed because the package was not properly designed or installed.

More recently, at least one major solar company has begun to use the service of two hire-purchase marketing agents. One of the hire-purchase groups (an Indian-owned firm called Wood Ventures) sends agents around the country approving credit and selling consumer items (furniture, televisions, radio-cassettes, sewing machines, etc.) on a hire-purchase basis. Its representatives make regular stops in towns and villages when teachers and civil servants receive their payslips, or when farmers are paid for their crops. They deduct repayments from government wages or co-operative payments on a 'check-off' system. The group added solar system packages to their 'catalogue' of products in 1995 and currently sells scores of systems each month, mostly based on amorphous products. Interest is above commercial lending rates (minimum 40%) and terms are usually about 18 months.

Solagen, a Nairobi-based retailer, offers a deposit system to customers who cannot raise enough funds to buy systems outright. Customers are given a price quote on a system and make a deposit with the company. They then arrange to pay the remaining amount over a fixed period. Solagen deposits customer payments in their account, and installs the system upon full payment. The advantages of this arrangement are that the customer avoids price increases over the deposit period, and Solagen is assured of a sale.

Present status

In the 1990s, a few major developments are changing the local solar market. Most important is the introduction of 12 Wp amorphous modules to the market from France, UK, Croatia, India and the USA. Amorphous modules are significantly less expensive than crystalline products and are available in smaller sizes (they are also more likely to fail). As much as 40% of the present Kenyan market is amorphous. This shift towards smaller, less expensive solar systems has occurred because demand for PV is high among income groups that cannot afford large systems. However, as is common with low-end products, there is a higher rate of failure of small (under 20 Wp) systems because of:

- under-sizing of systems
- poor quality equipment (batteries, modules, lamps)
- poor installation practice

The wide market for solar lanterns (less than 10 Wp) is recognised, but still small due to limited product choice, quality and availability.

The World Bank ESMAP sector is supporting a pilot project through Energy Alternatives Africa (EAA) and the Kenya Rural Enterprise Programme (K-REP). This will enable several hundred families to finance home lighting systems over about 18 months. Systems will be designed for each family, based on required system size and ability to pay. Through the K-REP credit programme, members of groups will guarantee each other's loans. EAA will insure that all systems are properly installed and designed, and that the local businesses installing the systems maintain them. An implementation manual is being prepared that will allow investors and/or energy supply companies to implement larger projects based on this model. The Kenya Rural Enterprise Programme is also providing small loans ($5-10,000) to Kenyan solar businesses to help them develop new products, energy services and marketing strategies.

The Kenyan Government has recently taken notice of PV's role in rural electrification. It reduced duty on PV modules to 10% and removed VAT (duties remain on regulators and batteries). The Government is also incorporating PV training into courses such as those offered at the Mombasa Polytechnic.

103

Kenyan Profile

Local Needs
Lights, TV, Radio

Technology
Solar Photovoltaics

Local Ability to Pay
Some wealth from cash crops
Increasing numbers of lower income households are able to pay

Technology Dissemination
Mainly commercial dealerships
Increasing movement towards NGOs

Financial disbursement mechanism
Traditionally up-front cash sales
Currently payment on installation is most popular
Trend is towards short term loans (3-18 months, through NGO structure

Typical Finance Terms
System cost: $500
Downpayment: 50% + 25% on installation
Repayment period: within 3 months
Interest rate: 0% if repayment complete within 3 months; 15% if repayment between 3 and 6 months

Fund establishment
Dealer's funds (possibly from Bank finance)

Lending policy
Lend to trusted customers with regular income
Repossession is an option if payment is not complete within 6 months

Lessons
Consumer can drive the PV market. This is now stimulating government interest.
The need for quality components, systems and service arrangements is paramount

VILLAGER POWER, SRI LANKA

Sri Lanka is an island with a population of around 17.4 million. Over three-quarters of this population is located in rural areas and about 10 million of these rural people do not have access to grid electricity.

Today, close to 5,000 PV systems are known to exist in Sri Lanka, installed through a combination of private sector, NGO and government efforts and using a number of different financing methods. Private sector initiatives based on commercial banks' or dealers' financing appear to be a viable mechanism provided there are sufficient incentives for the financiers and if the scheme is properly publicised. Highly subsidised government initiatives such as the Pansiyagam 1,000 Home Project using extended (e.g. twenty year) loan periods, minimal downpayments and low or zero interest charges, on the other hand, are not sustainable.

Sri Lankan experience indicates that finance schemes operating at a 'grass roots' level with some central supervision offer the best prospects for success. Two successful Sri Lankan models are the Sarvodaya Project and the Solanka Associates Project.

Sarvodaya Project

In 1992 the Solar Electric Light Fund (SELF), in association with Sri Lanka's largest NGO, the Sarvodaya Shramadana Movement, completed the introduction of solar electricity to 100 villages in rural Sri Lanka. 100 community centres, schools, shops, offices and Buddhist temples were equipped with 36 Wp solar electric lighting systems to demonstrate the value of solar power. In addition, ten villages instituted revolving loan schemes to finance a further 250 solar home systems.

The objective of the programme was to promote solar electrification as an alternative source of energy in rural communities of Sri Lanka, with the main emphasis of the project being placed on its finance programme.

The project was split into two phases:

- installation of PV systems for demonstration in 100 common places in different parts of Sri Lanka to increase public awareness of this new concept

- Sarvodaya Rural Technical Services (SRTS) to operate a revolving fund to cater to financing needs for the purchase of solar PV systems through the Rural Enterprises Programme (REP) of Sarvodaya

Phase 1

Within the Sarvodaya Society's country-wide infrastructure, 100 demonstration sites for phase 1 were identified. The only difficulty in the selection process was lack of access to Ceylon Electricity Board's plans for extending the grid which held the risk that power lines could come through some of these areas in the near future.

> The selection criteria for villages for phase 1 of the programme were:
>
> - villages would not have access to grid power in the next two years
>
> - villages are in the appropriate organisational stage of the Sarvodaya development process
>
> - villages are capably organised by the relevant staff to administer the credit programme

Once villages were identified, surveys of the villages were carried out by two Extension Officers who were recruited to implement the PV project. The survey obtained the following information:

- level of awareness about PV at the village level
- needs of the villagers
- general ability for potential users to pay

In early 1992, 16 solar technicians were trained in preparation for the implementation of the first phase. These technicians were drawn from skilled youth employed by SRTS.

The majority of the 100 demonstration installations were at village Buddhist temples,

Power & Sun Pvt Ltd (now the Solar Power and Light Company Ltd) was founded in 1986 by three Sri Lankan entrepreneurs following their return home after studies in Canada. The organisation, which was financed from the National Development Bank and Development Finance Corporation of Ceylon, was established to manufacture and market solar PV systems to households without access to the grid.

The company began laminating modules using imported cells, and turned out a simple line of domestic solar electric lighting kits in two sizes: 18 Wp to power three lights (8 Watt fluorescents) and a radio; and 36 Wp to power six lights, a 12V DC black and white television and a radio.

Marketing of the Suntec range was simplified by naming the systems '3L' and '6L', respectively, since the term 'watt peak' meant nothing to those at the village level. Power is rated in light hours, with two 'light hours' being equivalent to one 'TV hour' in terms of load.

looked after by the local Buddhist monk. The temples, which often serve as schools and community centres for evening education, were chosen to showcase solar lighting.

SELF provided the Suntec PV systems manufactured by Power & Sun Pvt Ltd. and SRTS conducted training courses in which 'showcase' systems were installed.

Phase 2

The Rural Enterprises Programme (REP), with technical advice from the Sarvodaya Rural Technical Services Programme (SRTS), introduced a loan scheme to members of the Sarvodaya Shramadana Societies to obtain PV systems for electricity facilities.

SELF organised a pilot revolving grass-roots credit scheme managed by Sarvodaya's Rural Enterprises Programme which administered the central solar fund. Sarvodaya's village credit co-operatives operated the programme locally, collecting monthly payments from

beneficiaries which were returned to the central fund.

To qualify for a loan, applicants had to be an active member of a Village Sarvodaya Shramadana Society; live in an area judged to be too far from the main electricity grid; be able to repay the loan instalment and interest without any constraints; and all repayments of previous loans should have been made on time.

150 systems were initially installed, with more being installed as payments replenished the revolving fund. Under the loan scheme two types of electricity supplies were introduced. System 1, the Suntec '6L', required a loan of US$450 and system 2, the '3L', a loan of US$300. Other equipment supplied for both systems was a charge indicator, a low voltage cut-out, a fused distribution board, battery, wires, fasteners and clips.

Each person taking out a loan had to provide personal security of two other members of the local Sarvodaya Society. If anyone defaults on their loan, then they and their guarantors will not be eligible to obtain any other loan from the society.

The systems were installed by SRTS free of charge and they also provide user training and maintenance.

The Sarvodaya Movement has now launched a national program with the goal of electrifying 1 million homes by the year 2005.

The Village Society was required to have the following qualifications to obtain credit:

- An active Shramadana Society by which Rural Enterprise Programme (REP) projects have already been started

- Accounting books of the society and savings/loan section should be satisfactorily maintained

- Repayments of the previous loans raised by the Society should have been made satisfactorily

Solanka Associates Project

Solanka Associates, an emerging NGO, has implemented two pilot projects in remote villages using seed funds raised outside Sri Lanka. With a total focus on solar PV, Solanka has provided an example of a successful model for the dissemination of solar PV.

Solanka trains and empowers the village community to administer the Programme. With the inherent difficulties of managing projects from cities, this model offers a lot of promise providing there is proper training and incentives to the operators at the village level. However, one of the barriers is that it takes commitment from a dedicated personnel and an organisation to expand this concept to other villages.

Lack of commercial incentives may be a barrier in expanding these programmes as promoters may not be able to survive on this work alone. Other problems lie in the area of securing funds to expand operations.

The Projects

The two projects, initiated with seed funds from the Solar Electric Light Fund and the Rotary Foundation, and supported by the provincial government, were implemented in the Morapathawa village (near the coastal town of Puttalam) with 84 homes being illuminated with solar PV and Thorawa, a village located 60 km away from Morapathawa, presently having 77 systems installed.

The project has an interesting feature where the 12 volt lamp units and the simple electronic controllers are manufactured at the village level. Also a village level battery repair unit has been initiated and defective cells can be replaced to lengthen the life of the battery.

The project is controlled through the Colombo based office, but most of the management functions have been transferred after training selected persons from the villages, so the day to day operations are now managed by the local leaders.

The repayments to date for both projects are 100% due to the grass roots level service that is provided to users. For instance, even when a battery fails, the user immediately gets a replacement while the old one is being repaired or serviced.

The Colombo based head office focuses on long term strategic planning and also imposes the accounting controls with audits of all accounting operations in the process of selecting recipients and collecting repayments.

A Commercial Approach for Solanka

With the community based experience behind it, Solanka now has set its course to take a commercial profit oriented approach. The promoter has realised the need for profit based operations as the NGO approach has draw-backs in terms of attracting funds. Funds are easier to access and the number of sources is bigger when projects are done on a commercial basis.

Sri Lankan Profile

Local Needs
Lighting for community centres, schools etc.
Lights, TV, radio for homes

Technology
Solar Photovoltaics

Local Ability to Pay
Survey undertaken to assess ability to pay

Technology Dissemination
Solar Electric Light Fund in association with local NGOs

Financial disbursement mechanism
Grass roots credit scheme managed by NGO at central level, with local branches responsible for overseeing loans to active members of the village society

Typical Finance Terms
System cost: $300 or $450
Downpayment: 10%
Repayment period: 5 years
Interest rate: 10%

Fund establishment
Revolving fund seeded with assistance of international benevolent foundations

Lending policy
The local branch responsible for loan disbursement must already be involved in rural enterprise programmes with well maintained accounts and good credit history.
Individuals are required to provide personal security from 2 other society members.
In the event of a default, defaulters and their guarantors become ineligible for further loans.

Lessons
Grass roots awareness of local client base, together with peer influence, can work as an RET dissemination method.

HEALTH CENTRES IN THE DEMOCRATIC REPUBLIC OF CONGO

In 1974 the World Health Organisation (WHO) initiated a global immunisation effort under the banner of the Expanded Programme on Immunisation (EPI). During the last twenty years, the EPI has made a valuable contribution to efforts aimed at controlling the spread of many potentially fatal diseases, but has had to contend with a high level of inoculation dose failures caused by the unreliability of cold storage facilities used to preserve the vaccines.

In rural areas which are remote from any electricity grid and often subject to fuel scarcity, photovoltaics offer one of the most reliable means of powering refrigeration systems for vaccine storage. The Democratic Republic of Congo (DRC), formerly Zaire, Africa's third largest country, covering an area of around 900,000 square miles has little electricity capacity outside its major cities, so in the mid 1980s, when the national ministry of health launched a programme to equip rural health centres with reliable vaccine storage and lighting facilities, PV was the preferred power source.

In 1984, the European Development Fund finalised funding for the world's largest rural health care project to utilise PV. 750 lighting and 100 solar refrigeration systems, (to be used for vaccine storage), were to be installed in health centres in DRC. The project covered five of DRC's twelve regions and not only included the testing of systems, but also installation, maintenance and the training of users. Installation was completed in 1990.

Despite the obvious implications for improved patient care and staff working conditions, the initial capital investment needed to equip the health centres with PV refrigeration and lighting emerged as a drawback. One way of sharing the initial cost of solar energy and of contributing to coverage of the recurrent costs of the EPI is to generate surplus solar electricity at the health centre, by installing a more powerful solar generator than is needed to run the refrigerator, and to sell the additional energy to the community, so as to produce income at the local level.

In 1990, the WHO/EPI and the Republic of Zaire initiated a pilot project in the Rural Health Zone of Nselo, Bas Zaire, to assess the feasibility of a such a scheme.

It was anticipated that the income generated by the sale of excess energy could be used, at each health centre or at the district level, to help recover the recurrent costs that are associated, in particular with the logistics of the EPI, such as:

- maintenance and repairs of cold chain equipment and vehicles used for immunisation operations

- fuel for these vehicles

- allowances for village health workers

- maintenance of health care premises and solar power systems

In terms of the initial investment required, the marginal cost of the oversized solar generator is low compared to that of solar refrigerator programmes. The transport, installation and maintenance costs are not increased by this additional system.

The major advantage for the investor is that a single initial injection of funds provides a health district with both a solar refrigerator and a permanent source of income that makes it possible to cover part of the operating costs.

Assessing local needs

The project was divided into two phases:

- a feasibility study was conducted during 1990 which was aimed at measuring the demand of the community for small quantities of electricity and defining the technical resources for meeting it

- a pilot project lasting for one year in which the health district was provided with energy distributors. The managers were briefed on the objectives of the project and the uses of the systems and of the income generated. This second phase was launched in August 1990.

109

Feasibility Study

The feasibility study, which examined the local market and evaluated the potential clientele, identified three major areas of interest to the community which were likely to generate substantial income and for which simple technology was available:

- recharging of motor-vehicle batteries - commonly used in rural areas for domestic purposes such as radios or television sets;

- rechargeable Ni-Cad dry cell batteries

- community television and video

Pilot Project Performance

After the study, a charger for recharging 12V car batteries, a charger for nickel cadmium batteries, and a TV with video player were installed at the Nselo Hospital. A 12V battery charger was also installed at a health centre 25 km from Nselo.

Since installation, the systems have functioned without major technical problems, and the demand has increased to be greater than the capability of the system to recharge batteries.

The mid-term evaluation conducted during 1991 showed that, with an investment (battery chargers, TV and video) that represented less than 13% of the total funds provided to equip the health centres with solar refrigerators and lighting systems, the scheme was able to generate more than 50% of the recurrent cost of the primary health care activities of the whole district.

The car battery charger was found to be a very simple and reliable technology, generating regular income for minimal managerial input and limited investment. During the evaluation period, the income resulting from payments for recharges outweighed the cost of providing the service by more than 50%. Break even was found to coincide with 5 batteries being brought to the recharging station every month. With on average 15 batteries being recharged every month, the profit potential was substantial.

Although the Ni-Cad charger performed well, some of the batteries initially introduced in the project were lost. Few villagers could afford the high capital costs of the batteries and they were therefore rented. However, despite a slow start on the rental of nickel cadmium batteries, there is now a faithful clientele who consider the batteries more durable and less expensive than the dry cells they can buy locally.

While the Ni-Cad recharging scheme could generate substantial income in principle, the arrangement has proved difficult to manage, and it was not seen as an income generating activity which should be repeated in future.

There were a few initial technical problems with the TV and video systems, but these were rapidly overcome. The scheme highlighted the need for robust equipment and strict technical specifications. Like the Ni-Cad scheme, the TV and video arrangement proved difficult to manage at first - largely because of the very high demand which can lead to dissatisfaction amongst a proportion of the users and which makes it difficult to control payment flows.

However, once the managerial problems were overcome, the TV service demonstrated its potential as an income generation scheme, with an additional benefit being the sociological possibilities that were offered in terms of popular health education and staff training through the showing of videos.

Overall the scheme offered considerable promise for replication in other regions, and indeed this pilot project was the basis for a PV lamp charging project in Senegal.

DRC Profile

Local Needs
Primary need for refrigeration and lighting for health care centres.
Secondary need for community TV and home power

Technology
Solar photovoltaics for direct use within health centres and to charge batteries for home power

Local Ability to Pay
Basis for purchase of charged battery is direct replacement of traditional kerosene fuel purchases or purchase of disposable batteries

Technology Dissemination
Systems installed at community health centres as part of a government health initiative supported by WHO. Health centre administers arrangements for communal TV access and battery recharging

Financial Disbursement Mechanism
N/A

Typical Finance Terms
N/A - cash payments for TV sessions and for recharged batteries

Fund Establishment
National governmental and international development agency support

Lending Policy
N/A - pilot programme, but local needs were assessed to confirm potential viability of scheme

Lessons
Offering community services - such as battery charging, and TV for educational or entertainment purposes - can be a valuable source of income for rural health centres (for example) and can have a positive impact on rural social development

INDONESIAN CO-OPERATIVE

In Indonesia approximately 25 million families still lack access to electricity. The government's efforts to electrify Indonesia have been hampered by the difficulties of supplying electricity to remote areas.

Indonesia first became involved with the development and application of PV systems in 1979, through the activities of the Agency for the Assessment and Application of Technology (BPP Teknologi).

However, the development of the Solar Home System (SHS) in Indonesia did not occur until 1987. The first SHS project was a result of co-operation between a Dutch PV company (R&S), BPPT and the Indonesian Ministry for Co-operatives and resulted in the installation of 80 SHS in Sukatani village in West Java. The Sukatani SHSs were introduced to demonstrate the feasibility of the PV solar concept for remote areas.

Installation of the SHS was undertaken by R&S with the help of a local co-operative which also provided a credit service, fee collection and simple maintenance. The seed money for the project was provided by R&S, the Netherlands Ministry of Foreign Affairs (DGIS) and BPPT. A downpayment and a monthly contribution was required from the users.

The President of Indonesia became enthusiastic about the Sukatani project and initiated the Presidential Assistance Project (Bantuan President Project or BANPRES) with interest free credit for 3,000 SHS. Monthly payments by villagers were used to pay back the loan component to the local co-operative at 0% interest. During the payback period of ten years the villagers use the SHS on a hire/purchase basis. After the loan component has been paid back, the ownership is transferred to the villagers. A small amount of the money paid back to the local co-operative is reserved for maintenance and repair. Through this system a financial fund is developed and at the same time it is used for other solar projects elsewhere.

One such scheme was implemented in the village of Lebak using a similar system design to those used in Sukatani. A revolving fund

> The SHS in Sukatani consists of the following components:
>
> - PV modules, each 40 Wp (type RSM 40)
> - charge controller
> - 12 V battery of 100 Ah
> - lighting fixtures (one 10 W, two 6 W)
> - accessories such as cables
> - instruction manual
>
> The following are typical loads to be connected to the PV SHS:
> fluorescent lamps (typically three)
> 12 V radio, 6 W
> 12 V black & white TV set, 12W

grant of US$50,000 was given by the North-Holland utility PEN. As part of the programme, several outdoor lights, public televisions and lighting systems for shops, workshops and public buildings were also installed.

In Lebak the village co-operative Cileles was chosen to deal with the administration of the project. The villagers pay their monthly contributions to the co-operative. On a national level, the project was managed in co-operation between the Dutch organisation R&S and BPP Teknologi.

The resources for maintenance of the systems were very limited. Some local people were trained to do basic maintenance such as topping up the batteries with water.

A complete SHS costs in the region of $500 and is therefore generally too expensive to be paid for in cash by the user. In Lebak the revolving fund system was chosen. The principle of a revolving fund is simple: the first batch of PV systems is bought and installed with money from the seed fund. The user then pays monthly instalments. With the money collected new SHSs can be financed, so the number of systems increases gradually.

Before the start of the PV project the villagers were mainly using oil lamps for lighting at a cost of $7 to $16 per month. To pay for their PV systems, the villagers had to pay $5 per month for ten years. After this period the PV systems officially belong to them.

CASE STUDIES

There have been a number of social benefits associated with the SHS programmes. These include improved conditions for children to study, better quality lighting, and the ability of households to enjoy electronic means of entertainment without the cost and potential environmental hazard of non-rechargeable batteries.

The installation of individual solar home systems has given the users an increased sense of responsibility for their systems. This has reinforced the importance of energy management within the home which abundant electricity from the grid tends to discourage.

By 1993 more than 10,000 SHS had been installed throughout Indonesia. The average investment came down to below US$400 and, with a downpayment of 35%, the SHS owners pay the equivalent of US$8 per month. It is estimated that there are now 20-30,000 solar home systems alone installed in Indonesia (IEA-PVPS, 1995b).

The Department of Co-operatives is assisting the Village Co-operatives (KUD) both financially and technically in setting up SHS projects. The remaining part of the funding is generated by Government and District budgets, and by customer payments.

The government has identified appropriate project financing as the single most important component for a steady growth of PV applications in Indonesia.

The SHSs are completely assembled in Indonesia. Apart from the modules, which are manufactured in the Netherlands, all components are also produced in Indonesia.

This positive situation in Indonesia is expected to grow due to the following reasons:

- increased dissemination of PV information through local governments, other government institutions, NGOs etc.

- increased competition for sale of systems, making for lower prices and better after-sale service

- proven reliability of applications in a number of sectors, leading to increased sales to these sectors, e.g. telecommunications

- experience in the field has shown that SHSs provide sufficient power for lighting, radios and television

113

Indonesian Profile

Local Needs
Lighting (+ TV and radio)

Technology
Solar photovoltaics

Local Ability to Pay
The avoided cost of oil lamp lighting exceeds typical monthly payments for PV system

Technology Dissemination
Led by a Dutch PV company and two Indonesian Ministries, undertaken by a local co-operative in association with the Dutch PV company

Financial Disbursement Mechanism
Villagers use the SHS under a hire purchase arrangement, based on interest free credit provided by the local co-operative

Typical Finance Terms
System cost: $400
Downpayment: 35%
Repayment period: 10 years
Interest rate: 0%

Fund Establishment
Via Presidential assistance, national government and Dutch utility support

Lending Policy
Information unavailable

Lessons
Individual SHSs give the user an increased sense of responsibility, and strengthen the importance of energy management within the home.

SHS SUCCESS IN THE DOMINICAN REPUBLIC

Due to a lack of available small, local financing schemes, the majority of PV systems in the Dominican Republic (DR) are paid for in cash by customers who have money and who have the ability to pay quickly for them. These include farmers, small ranchers, business owners and sewing businesses.

In a country whose per capita income is about twice the cost of a home lighting system, it is very hard for many people to raise the capital needed to buy such a system outright. Banks do not finance PV home systems. Between 5 and 10% of systems are bought using commercial financing through dealers. Less formal than banking finance, this short-term financing allows customers to make several payments over a 3-6 month period, paying an agreed interest rate to the dealer. As with community credit, this is based upon a trust relationship between the dealer and the customer.

Lack of longer term financing has been a factor limiting the wider dissemination of PV systems. In 1984 Richard Hansen founded Enersol, an NGO dedicated to providing solar electricity for rural households. He designed the 'SO-BASEC' programme - SOlar BASed rural Electrification Concept - to overcome the financing hurdle through the training of local technicians, the establishment of a service and sales centre and by helping local communities organise themselves into co-operatives running revolving funds. 5-10% of the systems installed through the Enersol programme are paid for on credit terms.

An Enersol study estimated that, with 15% interest rates on systems financed over three years, 2-30% of the unelectrified rural population would be able to buy their own systems. By down-sizing and reducing costs, this proportion could be raised.

In 1987 Enersol undertook a study to identify the potential demand for PV systems. It was discovered that 380,000, or 70% of rural homes had still not been electrified. According to consensus data, 66,000 of these had monthly incomes exceeding $200. This meant that these households would be able to make monthly payments of $8 over 7 years - the cost of buying a $500 home lighting system at 15% interest. This compares with other sources of non-grid electricity which have been purchased particularly by businesses and more affluent households, where generating sets typically cost about $90 per year per kW of operating capacity (1987), and campasenios (rural Dominicans) typically pay $10 per month for sources of light and electricity (kerosene and car batteries). Thousands of households use centrally-charged lead-acid batteries to power TV and radio (Hankins, 1993).

Enersol has worked with a number of credit groups, such as ADESOL (Asociación para el Desarrollo de Energia Solar - Solar Energy Development Association), which it helped form, and several NGOs. Typically, credit funds are set up by grants which provide replenishable start-up capital. Individuals who qualify for loans usually make a 25% downpayment, followed by principal and interest over 2-3 years.

Financing for PV home systems within the programme is facilitated through Enersol using a revolving fund, for which the seed money has come from international agencies. Loans are managed either through the NGO structure or, most commonly, the Community Association - which is made up of local PV users. Usually a downpayment of $50 is required, followed by monthly repayments equal to the amount of money saved each month by displacing the previous expenses for fossil fuels, battery charging and dry cells. Monthly instalments over the five year repayment period are typically between $8 and 12. The interest rate is 15%, which is seen as functional to protect credit depreciation, generate a small income for the local funding agency, and encourage people to pay off their loans so that the money can be re-lent for further purchase of PV systems.

An important factor in the success of the programme is the Solar Electric Service Enterprise which provides the equipment, spare parts and expertise needed at the local level to ensure that systems are available and continue to operate satisfactorily once installed. Technicians are also able to access credit facilities to buy modules through the revolving scheme of ADESOL. Some also have access to credit which they can then lend on to customers whom they think are credit-worthy and reliable.

In 1996, Enersol launched a three year project worth $1.5 million to install 1,000 stand-alone photovoltaic systems, as well as eight community water pumping systems powered by PV and wind, throughout the Dominican Republic. The project again incorporates consumer credit mechanisms to assist households in purchasing a solar home system (25 to 60 Wp) through a network of twenty solar-supply micro enterprises. On completion of the project, a revolving loan fund of $100,000 will remain for further end-user financing elsewhere.

Solar lighting systems are used for a number of income-generating activities. Most common are for country stores, which are now able to remain open during the evenings. This can also bring people to the stores for social activities, such as chatting and playing dominoes. Solar-powered lights allow women the opportunity to make additional income as seamstresses at night. Cafeterias use solar-powered DC blenders to make and sell fruit drinks.

Funding for different parts of the programme has come from a variety of sources. For instance, technician workshops have been funded by W Alton Jones Foundation and Citibank, and the Rockefeller Foundation has provided seed funding for new schemes. Enersol estimates that, based on an average loan of $400 and 15% interest rate, a revolving fund capitalised with $100,000 can finance approximately 800 PV systems over a 5 year period.

Dominican Republic Profile

Local Needs
Lighting (+ TV and radio)

Technology
Solar photovoltaics

Local Ability to Pay
Study indicated that income levels are sufficient to support monthly repayment charges - comparable to the price of purchasing 'traditional' fuel/power in the form of kerosene or centrally-charged batteries

Technology Dissemination
Through trained solar service and sales centres

Financial Disbursement Mechanism
Revolving fund with loan management undertaken by an NGO or local PV users' association. Sales and service technicians may also access credit through this structure and on-lend to customers

Typical Finance Terms
System cost: $500
Downpayment: $50
Repayment period: 5 years
Interest rate: 15%

Fund Establishment
Seeding usually with contribution from international agencies. May include a component from residual funds after completion of another revolving scheme

Lending Policy
Functional interest rate serves to encourage repayment. Distributors and local fund managers assess customers' credit-worthiness / reliability

Lessons
The total service approach - credit provision, systems and maintenance - is instrumental in the success of these programmes. Realistic interest rates protect against depreciation and provide support for the local agency

Appendix 1: Life Cycle Costing

Methodology

The methodology described here is based on a *discounted life-cycle cost analysis*. It is useful for calculating the least cost method of achieving a particular objective, e.g. a quantity of electricity generated, water pumped, or hours of lighting provided.

The inputs and outputs of the methodology are illustrated in Figure 1, overleaf. Information or assumptions are required to put numbers to:

- *system performance*: the output of the system depending on the natural resource available

- *cost data*: the up-front and future expenditure required

- *economic parameters*: the factors which dictate how the future costs can be expressed in today's money

These are combined mathematically to give economic indicators which summarise the cost-effectiveness of the system under consideration.

Life-cycle costing and discounting

In a life-cycle costing, the initial costs and all future costs for the entire operational life of a system are considered. The period for the analysis is normally the lifetime of the longest-lived system being compared.

To make a meaningful comparison, all future costs and benefits have to be *discounted* to their equivalent value in today's economy, called their *present worth* or PW. To achieve this, each future cost is multiplied by a *discount factor* calculated from the *discount rate*. The discount rate expresses how the value of money decreases the further into the future it is received.

For example, a discount rate of 10% per year would mean that in real terms it is equivalent for a customer to receive $100 now or $110 dollars in one year's time. Therefore a cost of $110 dollars one year from now has a present worth of $100. The discount rate can be viewed either as interest that would have to be paid on a loan, or the the interest that has been lost by spending capital rather than leaving it invested in the bank.

All calculations are done relative to general inflation, so that all costs are expressed in today's money.

117

FINANCING RENEWABLE ENERGY PROJECTS

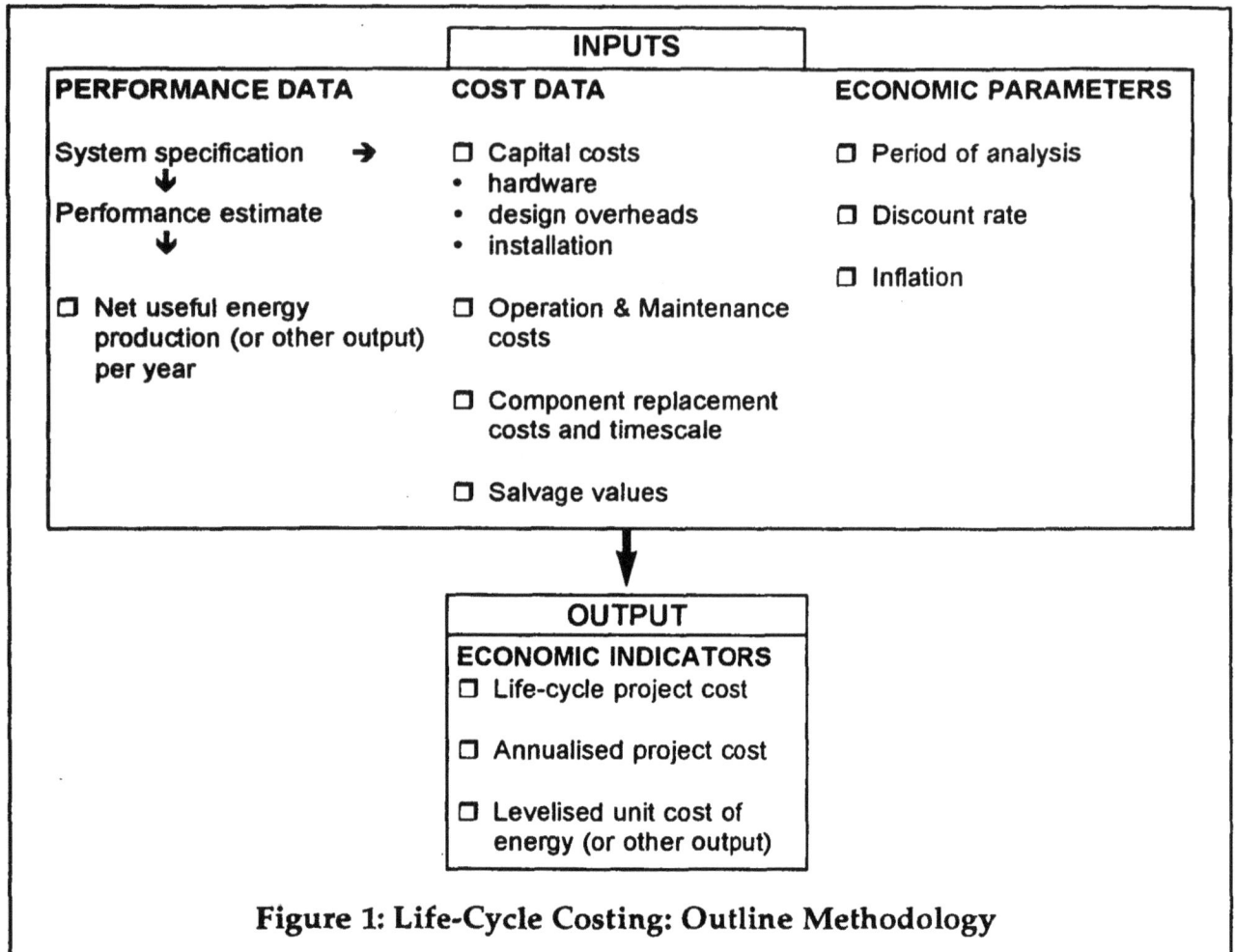

Figure 1: Life-Cycle Costing: Outline Methodology

Economic Parameters

The calculation of life-cycle costs requires values to be known for the following:

Period of Analysis (n)
Usually the lifetime of the longest-lived system under comparison.

Excess Inflation (h)
The rate of price increase of a component above (or below) general inflation (this is usually assumed to be zero).

Discount Rate (d)
The rate (relative to general inflation) at which money would increase in value if invested (typically 8-12%).

Calculation of Present Worth

There are two types of calculation that are used in life-cycle costing to express a future

cost at its present worth. The first is used to calculate the present worth of a single payment, say the replacement of a battery after five years. The second is used to calculate the total net present worth of a recurring cost, such as annual fuel or maintenance costs. In effect this is a quick method to calculate the sum of many discounted single payments over the analysis period.

Single payment

For a single future cost **Cr**, payable in **i** years time, the Present Worth is given by:

$$PW = Cr \times Pr$$

Where: $Pr = \left[\frac{(1+h)}{(1+d)}\right]^i$

and: **d** = discount rate

118

h = commodity-specific inflation rate (above general inflation)

i = year of incurred cost

Annual payment

For a payment **Ca** occurring annually for a period of n years the Present Worth is:

$$PW = Ca \times Pa$$

Where: $Pa = \left[\dfrac{a(1-a^n)}{(1-a)}\right]$

and: $a = \left[\dfrac{(1+h)}{(1+d)}\right]$

Life-Cycle Cost (LCC)

For each payment to be made during the lifetime of the system, the Present Worth can therefore be determined using the discount factors **Pr** and **Pa**. The sum of all PWs is the total life-cycle cost of the system:

$$LCC = C_{cap} + C_{rep} + C_{o\&m} + C_f - C_{sal}$$

Where:

C_{cap} = capital cost (the value of the total initial investment).

C_{rep} = discounted value of replacement costs of major items, e.g. batteries.

$C_{o\&m}$ = operation and maintenance costs, excluding fuel costs and major replacements costs, incurred over **n** years.

C_f = discounted value of annual fuel costs incurred over **n** years

C_{sal} = discounted salvage or re-sale value in year **i**.

Average values

Due to the difficulty in forecasting year-to-year changes in most of the analysis parameters, average or expected values have to be used. For example an average discount rate has to be applied, though the value of money may fluctuate at widely differing rates over the lifetime of the project. Similarly system efficiency will tend to degrade as components age and depending on the quality of maintenance work, whether it is a PV system or a diesel generator, so an average lifetime efficiency has to be assumed based on past experiences.

Economic Indicators

There are two ways that the life-cycle cost is commonly used to provide more intelligible expressions of system cost, namely:

● Annualised Life-Cycle Cost

● Levelised Energy Cost

Annualised Life-Cycle Cost (ALCC)

The Annualised Life-Cycle Cost is the LCC expressed in terms of a constant cost per year. It is the annual expenditure required to pay for the system over its lifetime and includes the cost of repayments on borrowed capital. It is therefore not simply the LCC divided by the number of years in the analysis, as this would take no account of the cost of borrowing money from a bank, or the opportunity cost of using private capital. The LCC must instead be divided by the factor **Pa**, found using the chosen discount rate, inflation rate, and the number of years equal to the analysis period. This is really the reverse process of discounting, and the result is expressed in $/year for each system. Therefore:

$$ALCC = \frac{LCC}{Pa(n)}$$

Levelised Energy Cost (LEC)

The levelised energy cost is probably the most useful figure for comparing two energy technologies. It expresses the average cost of generating each useful unit of energy during

119

the lifetime of a system. For example, if the system is generating electricity then it can be determined from the ALCC as follows:

$$LEC = \frac{ALCC\ (\$)}{Electricity\ supplied\ (kWh)}$$

The unit cost does not have to refer to electricity, but can alternatively be expressed in $/hour of lighting, or $/cubic metre of water pumped, etc. according to the system under analysis.

Example calculations

To illustrate the methodology, Figure 1 and Figure 2 lay out hypothetical life-cycle costing worksheets for a PV-Battery system and a 5kW diesel generator system respectively, following the methodology outlined above. Both systems are required to supply an average electrical load of 2.5 kWh/day for a lifetime of 20 years.

The values used for the different variables are believed to be typical for a developing country situation, but are purely illustrative. Calculations using these variables are detailed on the right-hand side of each worksheet.

It is important to remember that the calculations will be sensitive, to a greater or lesser extent, to any changes in the values of the input variables. It is therefore important to carry out sensitivity analyses to gauge the possible effects of these changes.

For instance, the PV system cost has a large component attributable to battery costs. If the battery cost in a particular location happened to be double that shown, or their lifetime only 3 years instead of 5, the overall life-cycle cost would rise considerably. Similarly, the diesel system has a high O&M cost. From past experience, this has been estimated as 15% per year of capital cost, but in reality will depend on local labour rates, cost and availability of spare parts etc. and might easily be half or double the figure shown.

APPENDIX 1 - LIFE CYCLE COSTING

ECONOMIC PARAMETERS				Calculations:
Period of analysis	n	20	years	
Discount Rate	d	0.1		
Inflation Rate	h	0		
Discount Factor	a	0.91		$a=(1+h)/(1+d)$
Annualisation Factor	$Pa(n)$	8.51		$Pa(n)=a(1-a^n)/(1-a)$

SYSTEM SPECIFICATION & PERFORMANCE				
LOAD				
Daily load	Ld	2.5	kWh/day	
Annual load	La	912.5	kWh/year	$La=365*Ld$
SOLAR RESOURCE				
Design Insolation	I	5.0	kWh/m2/day	
BATTERIES				
Battery Efficiency	$Ebat$	0.75		
Days of Battery Storage	$Sbat$	5	days	
Battery Size	$Gbat$	25.0	kWh	$Gbat=Sbat*Ld/0.5$
			(50% max.discharge)	
Battery Unit Price	$Kbat$	175.00	$/kWh	
Lifetime	$Nbat$	5	yrs	
PV SYSTEM				
Supply to batteries	Lpv	3.33	kWh/day	$Lpv=Ld/Ebat$
Array oversize factor	Xpv	1.40		oversized for low
				insolation periods
Array Size	Ppv	933	Wp	$Ppv=(1000*Lpv*Xpv)/I$
Module Unit Price	Kpv	5.00	$/Wp	
Lifetime	Npv	20	yrs	
STRUCTURE AND WIRING				
Support/Wiring Unit Price	$Kbos$	1.00	$/Wp	
Lifetime	$Nbos$	20	yrs	

COST DATA						
CAPITAL COSTS						
PV Array		Cpv		4666.67 $		$Cpv=Kpv*Ppv$
Batteries		$Cbat$		4375.00 $		$Cbat=Kbat*Gbat$
Support/Wiring		$Cbos$		933.33 $		$Cbos=Kbos*Ppv$
Installation (20%)		$Cins$		933.33 $		$Cins=0.2*Cpv$
SUB-TOTAL		$Ccap$			10908.33 $	
O&M Costs (2%)		Rom		93.33 $/year		$Rom=0.02*Cpv$
Life Cycle O&M costs		Com			794.60 $	$Com=Rom*Pa(n)$
REPLACEMENT COSTS						
Item	Yr [i]	Pr(i)	PW			$Pr(i)=x^i$
Battery	5	0.62	2716.53 $			$PW=Cbat*Pr(i)$
	10	0.39	1686.75 $			
	15	0.24	1047.34 $			
SUB-TOTAL	$Crep$				5450.62 $	
Salvage Value	$Csal$				-0.00 $	
	$Ctot$			TOTAL	17153.56 $	

ECONOMIC INDICATORS				
Total Life Cycle Cost	LCC		17153.56 $	$LCC=Ctot$
Annualised LCC	$ALCC$		2014.85 $/year	$ALCC=LCC/Pa(n)$
Levelised energy cost	Ke		2.21 $/kWh	$Ke=ALCC/La$

Figure 2: Example of Life-Cycle Costing: PV-Battery System

ECONOMIC PARAMETERS				Calculations:
Period of analysis	n	20	years	
Discount Rate	d	0.1		
Inflation Rate	h	0		
Discount Factor	a	0.91	/yr	$a = (1+h)/(1+d)$
Annualisation Factor	$Pa(n)$	8.51		$Pa(n) = a(1-a^n)/(1-a)$

SYSTEM SPECIFICATION & PERFORMANCE				
LOAD				
Daily	Ld	2.5	kWh/day	
Annual	La	912.5	kWh/year	$La = 365*Ld$
DIESEL SYSTEM				
Generator rated power	$Pgen$	5	kVA	
Generator cost	$Cgen$	5250.00	$	empirical: $Cgen = 4000 + 200*Pgen$
Av. Generator Efficiency	$Egen$	0.15		
Lifetime	$Ngen$	10	years	
FUEL				
Energy content	Gf	10.3	kWh/litre	
Fuel consumption	$Lgen$	1.62	litres/day	$Lgen = Ld/(Egen*Gf)$
Unit Fuel Price	Kf	0.80	$/litre	

COST DATA						
CAPITAL COSTS						
Diesel generator	$Cgen$		5250.00	$		$Cgen = Kgen*Pgen$
Installation (10%)	$Cins$		525.00	$		$Cins = 0.1*Cgen$
SUB-TOTAL	$Ccap$				5775.00 $	
O&M Costs (20%/yr)	Rom		1050.00	$/year		$Rom = 0.2*Cgen$
Life Cycle O&M costs	Com				8939.24 $	$Com = Rom*Pa(n)$
Fuel Cost at year 0	Rf		472.49	$/year		$Rf = Kf*Lgen*365$
Life Cycle fuel costs	Cf				4022.59 $	
REPLACEMENT COSTS						
Item	Yr [i]	Pr(i)	PW			$Pr(i) = x^i$
Diesel Generator	10	0.39	2226.51 $			$PW = Ccap*Pr(i)$
SUB-TOTAL $Crep$					2226.51 $	
Salvage Value $Csal$					-0.00 $	
$Ctot$				TOTAL	20963.34 $	

ECONOMIC INDICATORS				
Total Life Cycle Cost	LCC	20963.344	$	$LCC = Ctot$
Annualised LCC	$ALCC$	2462.35	$/year	$ALCC = LCC/Pa(n)$
Levelised energy cost	Ke	2.70	$/kWh	$Ke = ALCC/La$

Figure 3: Example of Life-Cycle Costing : 5kW diesel generator

Appendix 2: Financing model

Overview

The model simulates an organisation that sells or rents out PV home systems in a developing country. It is based on a straight-forward cash-flow analysis which helps the user to calculate the sales price of systems to recover costs and shows whether customer interest rates (and rental charges) are enough to keep the revolving fund sustainable.

Sales of systems on credit terms or through the rental arrangement require the securement and management of capital. In this case this is achieved through a revolving fund. Revenues into the revolving fund are counterbalanced by expenditure - for instance the purchase of new systems, repayment of debts and provision of service and maintenance. The difference between the two helps to determine the number of future financed systems sales (or rentals).

The user inputs data, including estimated sales volumes, and adjusts systems sales prices and interest charges until the fund shows a long-term sustainability.

Revolving fund management

The fund may be seeded by a combination of grants and debt financing. The performance of the fund is dependent upon the balance between revenues generated through systems sales and rentals, and the associated costs of providing the credit service.

The bulk of continuous revenue into the fund comes from customer repayments under the lease arrangement or payments made under the service or rental arrangements. Customers' initial downpayments (which also serve to limit the lending risk) also contribute to fund income.

The major uses of cash are the repayment of outstanding debts together with any accrued interest (when the seed is debt financed) and also the bulk purchase of new systems. The recurring costs of providing the credit service have been lumped together as either fixed, ie. independent of the number of systems financed, or variable (marginal), i.e. proportional to the number of systems financed. Fixed costs include salaries of management staff, supplies, marketing activities, utility bills etc. Marginal costs include the wages of payment collection and maintenance staff, and the purchase of materials and spare parts. The marginal maintenance cost is calculated from a separate spreadsheet, based on all labour, fuel, and any necessary component replacement costs incurred during the entire repayment period averaged for a single reporting period.

Presentation of data

For simplification, the model presents cash-flows on an annual basis. In practice, payment collections (and routine maintenance) would more likely be made on a monthly basis. The reason for using annual figures in the model is simply to reduce the number of recorded data points and make the long-term

Financing Programme Costs

> **Start-up costs**
> *(one-off charges at financing programme initialisation):*
> - purchase of land / buildings
> - purchase of office equipment
> - purchase of vehicles
>
> **Recurring fixed costs**
> *(independent of number of sales/rentals):*
> - salaries at HQ (e.g. of management, marketing and administrative staff)
> - marketing activities
> - building / equipment rentals
> - utility bills
> - office supplies
> - interest on outstanding debts
> - depreciation of fixed assets
>
> **Recurring variable costs**
> *(dependent on number of sales/rentals):*
> - materials (e.g. system bulk purchases, including import duty / sales tax where applicable)
> - component replacements
> - vehicles (fuel requirements thereof plus any new purchases)
> - salaries of field staff
> - cost of delayed or non-payments (usually a percentage of total sales)
> - corporation tax payable on profits
> - inflation

performance easier to discern. For practical implementation, payment collections should be regular enough that fund managers can gain an early indication of any potential repayment problems, but should not be so often that the associated costs - that is field staff wages, transportation costs (fuel, vehicle depreciation etc.) become overly significant.

Balancing revenues and costs

The base system cost is established by summing the costs of individual components, accounting for import duty where applicable. The sales price (i.e. the price which would be charged to customers purchasing the system with a one-off cash payment) is calculated by applying user-defined profit margin and sales tax onto the base cost.

The principal amount of the loan to each customer is calculated from the sales price less the downpayment. The calculation of the cus-

tomers' repayment annuity requires the user to specify repayment periods and the effective rate of interest charged to customers (see box for formula). Under the service arrangement, the anticipated customer payment is user-defined.

Fund income comes from customers' deposits and instalment payments. Income from deposits is proportional to the number of systems financed during the reporting period. Revenue from customer repayments is proportional to the total number of systems financed to date, less the total number of systems for which customers have completely repaid the original loan (only applicable to the lease arrangement) less the total number of customers lost through defaults. It is assumed that a fixed percentage of those customers still making payments under the finance arrangement will default each year.

Outgoings for each period comprise the total cost of purchasing any new systems (including any import duty payable), sales tax on systems sold, repayment of any outstanding debts (inclusive of interest), fixed costs as outlined above, and total variable costs which are proportional to the number of systems involved in the scheme during any given period (i.e. the total number of systems financed to date, less the total number of systems completely paid for, less the total default loss).

The revenue less the expenditure for each period constitutes the project cash-flow. This is added to the balance of funds from the previous period to determine the new fund balance from which, if sufficient funds are available, new systems will be financed.

> The annuity is calculated according to the formula:
> $$a = P \times \left[\frac{i(1+i)^n}{(1+i)^n - 1}\right]$$
> where:
> a = instalment payment (annuity)
> P = principal amount of loan
> i = interest rate per instalment period
> n = number of instalments

124

Assumptions

- All values are in real terms. It is assumed that system costs, labour, fuel, vehicles, office equipment etc. all rise in line with inflation.

- Normally no systems will be financed during any period if the available funds from the previous reporting period are zero (or negative).

- The demand for the systems always matches that specified by the user (i.e. there is no prospect of there being no customers to partake in the financing pro-gramme unless the user intervenes to set sales figures for any period to zero).

- All equipment costs (purchase of office equipment, vehicles etc.) are borne during the project start-up phase. In practice there would be some periodic replacement of worn-out equipment and purchase of new equipment to accommodate programme expansion.

- Systems reclaimed from defaulters are not re-used. In practice such a situation is unlikely as there would then be no benefit for the financing organisation in accepting the system as collateral.

Glossary

ASEAN	Association of South-East Asian Nations.
ASTAE	Asia Alternative Energy Unit, of the World Bank.
Alternating Current (AC)	Electric current in which the direction of the flow is reversed at frequent intervals. The conventional grid supply is AC with an alternating frequency of 60 Hz.
Amortisation	The liquidation of a debt, usually by periodic payments.
Amorphous	The condition of a solid in which the atoms are not arranged in an orderly pattern; not crystalline.
Amp	The basic unit of electric current (*the constant current that, when maintained in two parallel conductors of infinite length and negligible cross-section placed 1 metre apart in a vacuum, produces a force of 2×10^7 newton metres between them*).
Ampere-hour (Amp-hour or Ah)	A measure of electrical charge, equalling the quantity of electricity flowing in one hour past any point of a circuit. Battery capacity is measured in amp-hours.
Annuity	A stream of equal cash flows from period to period.
Array (photovoltaic)	A group of photovoltaic modules wired together to produce a specific amount of power. Array size can range from one to hundreds of modules, depending on how much power will be needed.

127

Asset	Anything of value owned by an organisation or individual.
Average collection period	A financial measurement showing number of days the average receivable is outstanding.
Average cost of debt	A financial measurement showing the ratio of interest charges to total debt.
BANPRES	Indonesian Solar Home Project, financed by the Presidential aid programme.
BPPT	Indonesian Agency for the Assessment and Application of Technology.
BRI	Bank Rakyat Indonesia.
Balance of payments	A summary of all of the payments by residents of one country (A) to other countries with payments received by that country (A) from each other country. If more money is received than paid then there is a favourable (surplus) balance of trade. If more money is paid than received then there is an unfavourable (deficit) balance of trade.
Balance of system (BOS)	The parts of a photovoltaic system other than the PV array itself, e.g. support structures, wiring, power conditioning etc.
Benefit-cost ratio	A tool used in capital budgeting to measure a profitability index.
Biomass	Organic waste from vegetation, animals and humans.
Biogas	Gas from biomass substances.
Break-even pricing	The point at which revenues are matched with costs. Any number of units beyond the break-even point within current production capability will contribute to profit. Conversely, if total number of units sold is below the break-even point a loss will result.
Broker	A party who does not take title to goods or property but brings buyers and sellers together and assists in the negotiations.
Budget	A financial plan which is used to measure performance. Budget variance may be favourable if costs are less than forecast, or unfavourable if costs are higher than those allocated.

Bureaucracy	A theoretical framework of authority designed for a rational organisation to perform tasks with high efficiency.
CASE	Centre for the Application of Solar Energy, Perth, Australia.
CDER	Centre de Développment des Energies Renouvelables - Morocco.
CILSS	Comité Inter-Etats pour la Lutte contre la Secheresse dans le Sahel - Agency combating desertification in the Sahel region of west Africa.
CFL	Compact fluorescent lights.
Capacity Factor	Ratio of energy output per year to the maximum output if the system runs at full-rated capacity all year around.
Capital	Money, etc. with which a company (or individual) starts business. Also accumulated wealth, typically of a business, in terms of stock, equipment and other assets.
Cash flow	The net profit after taxes plus non-cash expenses such as depreciation.
Charge controller	A component of a photovoltaic system that controls the flow of current to and from the battery to protect the battery from over-charge and over-discharge. The charge controller may also indicate the system operational status.
Collateral	Assets that are pledged to secure a loan, thereby reducing risk to the lender.
Commercial finance companies	These organisations provide short-term loans primarily for financing tangible assets (e.g. physical goods such as equipment etc.).
Compound interest	Interest continues to be calculated on accumulated interest in addition to the principal balance of a loan.
Conversion efficiency	The ratio of the electrical energy produced by a photovoltaic cell (or module) to the energy from the sunlight incident on the cell (or module). This is usually quoted for standard test conditions (STC).
Cookstove	Traditional small domestic biomass cooking stove. Usually portable.

129

Cost/benefit analysis	A comparison of the costs to be incurred with the benefits to be derived. The results are used as part of a decision-making process.
Cost of capital	The interest expense incurred by an organisation for financing business activities.
Credit policy analysis	A review of all the variables in connection with credit granting aid. Assumption is that when credit standards are eased sales will increase and if standards are tightened sales will decrease.
Crystalline	The condition of a solid where the atoms are arranged in an ordered pattern.
Cycle life (Battery)	The number of times the battery can be charged and discharged to a specific level before failure.
DANIDA	Danish Development Agency.
DGIS	Netherlands' Agency for Development Co-operation.
DC	Direct Current.
DRC	Democratic Republic of Congo (formerly Zaire).
Debt maturity	The date upon which the final payment of a debt is due.
Debt-to-asset ratio	This is measured by dividing the total liabilities by total assets.
Debt-to-equity ratio	This ratio is measured by dividing total liabilities by total stockholders' equity.
Deferred annuity	This is an annuity where payments are due at the beginning of each period.
Depreciation	The reduction in market value of an asset over its useful life.
Diffuse radiation	Solar radiation scattered by the atmosphere.
Direct costs	These are the costs incurred in the manufacturing process that relate directly to the output of a product or service.
Direct Current (DC)	Electric current in which electrons flow in only one direction. This is the current flow produced by a solar system. To be used for

typical 110-volt or 230-volt household appliances, it must be converted into alternating current.

Direct radiation	Solar radiation transmitted directly through the atmosphere.
Discount factor	The calculation for a certain interest rate with the number of time periods considered.
Discounted cash flow	The amount of present value to forecast future cash flows.
Distribution channel	A marketing system that moves products from the manufacturer to distributor to ultimate consumer or industrial user. The channel may be short (manufacturer to consumer) or have several intermediaries.
ECRE	Export Council for Renewable Energy.
EDF	European Development Fund.
EPI	Expanded Programme on Immunisation, the WHO global immunisation effort.
ESCO	Energy Service Company.
ESMAP	Energy Sector Management Assistance Programme. UNDP/World Bank technical assistance unit for the energy sector.
Economic analysis	An examination of the cost/benefit relationship relating to investment decision-making.
Efficiency (of a solar cell or module)	The ratio of electric energy produced to the amount of solar energy incident on the cell or module. Typical crystalline solar modules are about 12-14% efficient - they convert about 12-14% of the light energy they receive into electricity.
FINESSE	Financing Energy Services for Small-Scale Energy-users. Programme administered by the UNDP.
FONDEM	Fondation Energies pour le Monde - French development NGO.
FWWB/India	Friends of Women's Banking - India.
Financial intermediaries	These organisations include banks, insurance companies, factoring companies, and others that arrange financing for organisations and protect assets of transacting organisations.

Fixed costs	Generally, these costs are not affected by the number of units manufactured and sold.
Foreign exchange market	This market engages in the trading of a currency of one currency for that of another.
GAP	Global Approval Programme.
GDP	Gross Domestic Product.
GEF	Global Environment Facility. This is jointly administered by the World Bank, UNDP and UNEP.
GTZ	Deutsche Gesellschaft für Technische Zusammenarbeit, the German government department for international development.
Global irradiance	The total irradiance (sunlight intensity) falling on a surface; the sum of the direct and diffuse irradiance.
Grid-connected	A (photovoltaic) system that is connected to a centralised electrical power network.
Grid-extension	Extension of the electric grid.
Gross National Product (GNP)	The absolute value of all goods and services produced in a country.
Hard currency	A strong currency that may be easily converted into another currency. For example, Yen may be converted into US Dollars without difficulty.
Hard loans	Loans that require payment in a hard currency and at current market interest rates.
Head	The height through which water must be pumped, or the height of the water column provided by a raised tank.
Hybrid system	A power system consisting of two or more power generating sub-systems (e.g. the combination of a wind turbine or diesel generator and a photovoltaic system.
IBRD	International Bank for Reconstruction and Development, part of the World Bank Group. Lends to governments.
IDA	International Development Association. Part of the World Bank Group, lends to Least Developed Countries.

132

IFREE	Institute for Renewable Energy and Energy Efficiency.
IFC	International Finance Corporation. Part of the World Bank Group. Lends to the private sector in developing countries.
ISO	International Standards Organisation.
IREDA	Indian Renewable Energy Development Agency.
Improved cookstove	Traditional cookstove which has been made more efficient.
Indirect expense	Costs that are not easily traceable to one department of an enterprise. These costs may require an allocation among several departments. (Also referred to as common costs.)
Inflation	An economic condition that results in overall general price increases, thereby causing higher costs. This could result in a spiral that keeps generating higher and higher prices.
Insolation	Common term used to describe the amount of solar energy received on a surface. Usually expressed in Watts per square metre (W/m²), but also expressed on a daily basis as Watt-hours per square metre per day (Wh/m²/day).
Intangible asset	An asset without a physical form, but which allows rights to current and/or future benefits. Examples include: goodwill, patents, and copyrights.
Interest	The amount charged (interest revenue) by the lender for the use of the principal amount and the amount paid (interest expense) by the borrower for its use.
Intermediary	A go-between, a person or organisation which acts on behalf of another organisation to liaise with or provide a service to a third party.
Internal rate of return	The discount rate that equates the cost of the investment with the present value of future cash flows.
Inverter	A device that converts direct current (DC) into alternating current (AC) electricity.

Irradiation	The amount of solar energy received on a surface (kWh/m^2).
K-REP	Kenya Rural Enterprise Programme.
KUPEDES	Kredit Umum Pedesaan; BRI programme of general rural credit offered at commercial rates.
Kilowatt (kW)	Unit of power equal to 1,000 Watts.
Kilowatt hour (kWh)	1,000 Watt-hours of energy.
Kilowatt peak (kWp)	Power output of a photovoltaic module under standard test conditions.
LDCs	Least Developed Countries.
Lease	A contract whereby the lessee pays for the rights to use certain assets for a specified period of time. The party who owns the assets is the lessor.
Liabilities	Amounts that are owed to creditors.
Life-cycle cost (LCC) analysis	A form of economic analysis to calculate the total expected costs of ownership over the lifespan of the system. LCC analysis allows a direct comparison of the costs of alternative energy systems, such as photovoltaics, fossil fuel generators, or extending utility power lines.
Load	In an electrical circuit, any device or appliance that uses power (such as a light bulb or water pump).
MIGA	Multilateral Investment Guarantee Agency. Part of the World Bank Group, assists developing countries to attract foreign investments.
MSF	Médicins Sans Frontières - French medical NGO.
Maintenance costs	Any costs incurred in the upkeep of the system. These costs may include replacement or repair of components.
Market interest rate	The interest rate at which investors are willing to risk their money. This is sometimes referred to as the effective rate of interest.

Micro-hydro	Turbine which converts water pressure into mechanical shaft power. Typically less than 300 kW in size.
Mini-hydro	Turbine which converts water pressure into mechanical shaft power. Typically less than 2,000 kW in size.
Mortgage	A pledge of a certain asset that must be forfeited if the borrower does not repay a debt as set forth in the terms and conditions of an agreement.
NGO	Non-Government Organisation.
NRECA	National Rural Electrification Co-operative Association.
Net income (net loss)	The variance between net revenues and total expenses for a given accounting period.
Not-for-profit or NGO	An organisation that is structured to provide services and/or products to populations with special needs, without making profits for itself.
OECD	Organisation for Economic Co-operation and Development.
Operating costs	The cost of using a system. For fuel-based systems these costs include all fuel costs over the system's lifetime.
Opportunity cost	The return of the best alternative investment that will be foregone because limited resources are channelled elsewhere.
Orientation	The direction of the PV system surface.
Orientation angle	The angle between the normal to the PV system surface and the equator.
PPER	Programme of Rural Pre-Electrification.
PRONASOL	Programa Nacional de Solidaridad, Mexico.
PV	Common abbreviation for photovoltaic.
PVMTI	PV Market Transformation Initiative.
Peak Watts (Wp)	The maximum power (in Watts) a solar array will produce on a clear, sunny day while the array is in full sunlight and operating at 25°C. Actual wattage at higher temperatures is usually somewhat lower.

135

Payback period	The amount of time required for a project to return the capital investment made for it. No allowance is made for the time value of money.
Photovoltaic cell	Semiconductor device that converts light to electricity using the photovoltaic effect.
Photovoltaic module	A number of photovoltaic cells electrically interconnected and encapsulated into a sealed unit of a size convenient for handling, shipping and assembling into an array. Also called a 'panel'.
Photovoltaic system	A complete set of interconnected components for converting sunlight into electricity by the photovoltaic process, including array, balance of system components, and the load.
Power	The rate at which energy is consumed or generated. Power is measured in Watts.
Power conditioner	The electrical equipment used to convert power from a photovoltaic array into a form suitable for subsequent use. Also a loose collective term for inverter, transformer, voltage regulator, and other power controls.
Pre-electrification	Where full-scale electrification has not occurred for technical or economic reasons, pre-electrification consists of providing communities with minimal qualities of electricity to satisfy basic needs (e.g. for lights, radio, TV).
Principal	Capital sum lent or invested.
RAPS	Remote Areas Power Supply.
RETs	Renewable Energy Technologies.
ROSCA	Rotating Savings and Credit Association.
Renewable energy	Flows of energy that are regenerative or virtually inexhaustible. Most commonly includes solar (electricity and thermal), biomass, biogas, geothermal, wind, tidal, wave and hydro power sources.
Return on Investment	A ratio of income to assets or net worth, that may be calculated for an enterprise, or a product.

Revolving fund	A fund or scheme where the running costs are met by the continuous return of loan capital with interest.
SADC	Southern African Development Community.
SEI	Stockholm Environment Institution.
SELF	Solar Electric Light Fund.
SEP	Special Energy Project.
SHS(s)	Solar Home System(s). Small PV-powered system, usually consisting of a few lights and TV.
SO-BASEC	SOlar-BAsed Rural Electrification Concept.
Secured debt	A kind of debt that has been collateralised. Typically, this debt offers some degree of protection to a certain class of bondholders.
Seed funding	Capital provided to start a project or fund. No repayment necessary.
Silicon	A non-metallic element that, when specially treated, is sensitive to light and capable of transforming light into electricity. Silicon is the basic material of sand, and is the raw material used to manufacture most photovoltaic cells.
Small hydro turbine	Turbine which converts water pressure into mechanical shaft power. Typically less than 10,000 kW in size. (Mini-hydro is usually less than 2,000 kW and micro-hydro less than 300 kW.)
Soft loan	Loan issued on more favourable terms than commercial lending rates (e.g. lower interest rates and longer term).
Solar cooker	Cooker which uses the heat of the sun to cook the food.
Solar dryer	Technology which uses the heat from the sun to dry food (usually).
Solar still	Equipment which produces clean water through distillation, using the heat from the sun.
Solar thermal electric systems	Equipment which uses direct radiation to heat the working fluid of a heat engine connected to a generator.

137

Solar water heater (SWH)	Equipment which uses the heat from the sun to warm water.
Stand-alone systems	Power systems which are not connected to the electric grid. Most stand-alone systems include some type of energy storage, such as batteries.
TSECS	Tuvalu Solar Electric Co-operative Society.
Tariff	A tax on imports that is required by the government. The prime purpose of a tariff is to raise the price of imported products, thereby giving an advantage to domestically produced goods. Another purpose of a tariff is to raise revenues for the government.
Term loan	Generally, a bank loan would be considered long term if its maturity date is more than one year. A term loan is usually repaid with an amortisation schedule with monthly or quarterly instalments.
UNCTAD	United Nations Conference on Trade and Development.
UNDP	United Nations Development Programme.
UNEP	United Nations Environment Programme.
UNESCO	United Nations Educational, Scientific and Cultural Organisation.
UNICEF	United Nations Children's Fund.
UNIDO	United Nations Industrial Development Organisation.
USAID	US Agency for International Development.
Unsecured debt	A loan with no pledged assets as collateral. However, the bondholder does maintain a general claim against the enterprise rather than a claim against certain assets. Unsecured bonds are referred to as debentures.
Venture capital	Capital investment that is targeted for small or start-up enterprises that are in need of a infusion of capital for growth. There is a high risk associated with venture capital. Therefore, the venture capitalist usually offers equity capital and may require a major portion of stock ownership.

Voltage	A measure of the force or 'push' given to the electrons in an electrical circuit; a measure of electric potential. One volt produces one amp of current when acting against the resistance of one ohm.
WHO	World Health Organisation.
WWB	Women's World Banking.
Watt (W)	Unit of power.
Watt-hour (Wh)	Unit of energy; one Wh is consumed when one W of power is used for a period of one hour.
Watt peak (Wp)	The power output of a PV module under standard test conditions. It is the approximate amount of power that a photovoltaic module or array produces at noon on a clear day, such that the irradiance is equal to 1,000 W per square metre when the device directly faces the sun.
Wind pump	Mechanical device which uses the power of the wind to pump water.
Wind turbine	Electrical device which uses the power of the wind to generate electricity.
Working capital	The sum of current assets minus current liabilities. Working capital is a measurement of an enterprise to meet its short-term debt with its current assets.

Bibliography

Abugre, Charles (1993), 'When Credit is not due - financial services by NGOs in Africa', Small Enterprise Development, vol. 4 no. 4, IT Publications, London, UK.

Aguilera, J and Lorenzo, E (1995), Lessons Learned from a PV Rural Electrification at the Bolivian High Plato, 13th European Photovoltaic Solar Energy Conference, Nice.

Akura, Terubentau (1996), The Solar Photovoltaic Projects in Kirabati, Regional Workshop on Solar Power Generation Using Photovoltaic Technology, Asian Development Bank, Manila.

Anderson, D & Ahmed, K (1995), The Case for Solar Energy Investments, World Bank Technical Paper No. 279, Washington DC.

Anderson, Denis and Williams, Robert (1993) The Cost-Effectiveness of GEF Projects, Global Environment Facility, Washington DC.

Bakthavatsalam, V (1995), Renewable Energy Project Implementation in India, Workshop Proceedings on Financing the Development and Deployment of Renewable Energy Technologies, Oakridge Institute for Science and Education for the US Department of Energy, contract DE-AC05-760R00033.

Barlow, R, Derrick, A & McNelis, B (1993), Solar Pumping: An Introduction and Update on the Technology, Performance, Costs and Economics, IT Publications, London.

Barlow, R., Bokalders, V., Crick, F., Derrick, A., Fraenkel, P. (1993), Windpumps: A Guide for Development Workers, IT Publications, London.

Berdai, M, Butin, V, Huacuz, J M, Bakker, P & Anand, I S (1996), Decentralised Rural Electrification Organisational Aspects, Energy Development Group, Bulletin No. 5, ADEME, Paris.

Berkovski, Boris (May 1995), 'World Solar Summit Process', Workshop Proceedings on Financing the Development and Deployment of Renewable Energy Technologies, Oakridge Institute for Science and Education for the US Department of Energy, contract DE-AC05-760R00033.

Bezdek, Roger H, (July 1981), Economic and Institutional Barriers to Renewable Energy Technologies, Conference on Financial Issues for International Renewable Energy Opportunities, Virginia, USA.

Biermann, E, Corvinus, F, Herberg, T & Höfling, H (1995), Basic Electrification for Rural Households, GTZ, Germany.

Böhnke, Heinz-Wolfgang (1996), Photovoltaic Electrification, The Experience of GTZ in Disseminating PV-Technology in The Philippines, Regional Workshop on Solar Power Generation Using Photovoltaic Technology, Asian Development Bank, Manila.

Borg, Nils (ed) (1996), *International Association of Energy-Efficient Lighting Newsletter*, Issue no. 14, vol. 5, Stockholm.

Bromley, Anthony (May 1995), *'UNIDO's Programme for Promoting Solar Energy Applications'*, *Workshop Proceedings on Financing the Development and Deployment of Renewable Energy Technologies*, Oakridge Institute for Science and Education for the US Department of Energy, contract DE-AC05-760R00033.

Bruno, Eva-Elena (1996), *International Energy Agency Implementing Agreement on Photovoltaic Power Systems, Annual Report 1995*, ENEL, Rome.

Cabraal, Anil, Cosgrove-Davies, Mac & Schaeffer, Loretta (1996) *Best Practices for Photovoltaic Household Electrification Programs, Lessons from Experiences in Selected Countries*, The World Bank, Washington DC.

Castello, Carlos, Strearns, Katherine and Christenson, Robert Peck (1991), *Exposing Interest Rates: their true significance for micro-entrepreneurs and credit programs*, ACCION International, Discussion Paper Document No. 6, Washington DC.

Chadraa, B, Derrick, A, Purevdorge, G (1994), *Photovoltaics for Household Energy in Mongolia: Experience and Potential*, IEEE First World Conference on Photovoltaic Energy Conversion, Hawaii.

Christen, R P (1989) *What Microenterprise Credit Programs can learn from the moneylenders*, ACCION International, Washington DC.

Climate Network Europe (1996), *A NGO Guide to the Global Environment Facility*, Geneva.

Cochran, Jacquie (1995), *The UPVG Record*, Utility Photovoltaic Group, Washington DC.

Conway, James and Wade, Herbert (December 1994), *Photovoltaic Electrification in Rural Tuvalu*, Forum on Renewable Energy, a Key to Sustainable Growth in the Asia-Pacific Region, Hawaii.

Cross, Bruce (ed) (1995), *The World Directory of Renewable Energy Suppliers and Services*, James and James Science Publishers, London.

Davis, Mark and Dickson, Bruce, (June 1994), *PV dissemination to rural households*, The African Sun, Solar Energy Society of South Africa, Pretoria, South Africa.

deLucia, Russell J and International Fund for Renewable Energy & Energy Efficiency (1995), *Financing Renewable Energy Projects: Issues, Options and Innovations for Asia and Beyond*, Washington DC.

Department of Energy, (1994), *Renewable Energy Power Program, guidelines*, DoE, Manila, Philippines.

Derrick, A, Francis, C & Bokalders, V (1991), *Solar Photovoltaic Products: A guide for development workers*, IT Publications, London.

Derrick, A, Durand, J M and Hart, T J (1991), *Solar-powered Vaccine Refrigerators: Results of Concerted Action on Infrastructure Support*, 10th EC PV Solar Energy Conference, Lisbon, April.

Durand, J M and Zaffran, M (April 1991), *Sale of Excess Solar Energy: a contribution to the recurrent costs of Immunisation programs*, 10th European Photovoltaic Solar Energy Conference, Lisbon, Portugal.

Environment Matters at the World Bank (1996), World Bank Environmental Projects, The World Bank, Washington DC.

ECSCR (June 1994), *Second International Solar Cooker Test: Summary of Results (2nd Edition)*, European Committee for Solar Cooker Research, Lodève.

EUREC Agency (1996), *The Future for Renewable Energy*, James & James Science Publishers, London.

European Solar Industry Federation (1996), *The Sun in Action: The Solar Thermal Market*, European Commission, Contract No. 4.1030/E/94-003.

European Photovoltaic Industries Association (1996), *Photovoltaics in 2010. The World PV Market to 2010*, European Commission, Altener Study 4030.93.06.

Evaluation of PV-Household Rural Electrification project (1993), SCNCER Technical Information Service, ASEAN Sub-Committee on Non-Conventional Energy Research, Bangkok.

FDC (1992), *Banking with the Poor*, Second Regional Workshop on Banking with the Poor, Kuala Lumpur, The Foundation for Development Cooperation, Brisbane.

Fraenkel, P, Paish, O, Bokalders, V, Harvey, A, Brown, A & Edwards, R (1991), *Micro-Hydro Power: A guide for development workers*, IT Publications, London.

Freeling, Robert A (September/October 1995), '*Solar Electric Light Fund: Bringing Power to the People*', Solar Today.

Gabler, H & Beurskens, J (September 1996), *Rural Electrification with Photovoltaics*, EuroSun '96, Freiburg, Germany.

Germidis, Dimitri, Kessler, Denis, & Meghir, Rachel (1991), *Financial Systems and Development: What role for the formal and informal financial sectors?* OECD, Paris.

Gianto, Dr H, *Evaluation of PV Household Rural Electrification Project*, Report by the Ministry of Co-operatives and Small Enterprise Promotion, Indonesia.

Global Environment Facility (1992a), *Zimbabwe, Photovoltaics for Household and Community Use*, Project Document, World Bank, Washington DC.

Global Environment Facility (1992b), *India Alternate Energy Project, Project Document*, World Bank, Washington DC.

Go Between (December - January 1996/97), United Nations Non Government Liaison Service, Geneva.

Gregory, Jenniy (1994), *Financing Mechanisms for Solar Energy Technologies*, AFREPREN & Stockholm Environment Institute, Nairobi.

Gregory, J A & McNelis, B (December 1994), *Non-technical barriers to the commercialisation of PV in developing countries*, First World Conference on Photovoltaic Energy Conversion, Hawaii.

Gregory, J A, Bahaj, A S & Stainton, R S (April 1993), *Barriers to the development of PV*, 12th European Photovoltaic Solar Energy Conference, Amsterdam.

Gullberg, Monica (1997, forthcoming): *The first rural electrification co-operative in Tanzania - UECCO*, Stockholm Environment Institute.

Hankins, Mark (1995), *Solar Electric Systems for Africa*, Commonwealth Science Council, London.

Hankins, Mark (1993), *Solar Rural Electrification in the Developing World*, Solar Electric Light Fund, Washington DC.

Hansen, Richard (21-25 October 1991), *Making the Investment: Innovative ideas in alternative energy policy, financing and services*, Financing of Energy Services for Small-scale Energy Users Workshop, Kuala Lumpur, Malaysia.

Harper, Malcolm (ed) (1997), *Partnership Financing for Small Enterprise, Some Lessons from Islamic credit systems*, IT Publications, London.

Hass and Bender (Spring 1996), The Problem of Attracting Long-Term Debt for Privately-Financed Infrastructure, *The Journal of Project Finance*, Vol. 2, No. 1.

Hilhorst, Thea & Oppenoorth, Harry (1992) *Financing Women's Enterprise*, IT Publications, London.

Hoeke, P (1993), *Solar Electricity in Lebak, Indonesia*, Perpetuum Mobile of the 21st Century, December.

Hoffman, R T and McNelis, B (1986), *Building the renewable energy market in developing countries*, European Commission, Directorate General for Development, Brussels, February.

Holt, Sharon and Ribe, Helena (1990), *Developing Financial Institutions for the Poor and Reducing Barriers to Access for Women*, World Bank Discussion Paper 117, World Bank, Washington DC.

Huacaz, Jorge and Martinez, Ana M (1992), *PV for Rural Electrification: Early Mexican experience*, Advanced Technology Assessment System Issue 8, Prospects for Photovoltaics, United Nations, New York.

Huacaz, Jorge and Mulas del Pozo, Pablo, (December 1994), *Mexico Renewable Energy Programs and Prospects*, Forum on Renewable Energy, a Key to Sustainable Growth in the Asia-Pacific Region, Hawaii.

Hulme, David (1993), 'Replicating Finance Programs in Malawi and Malaysia', *Small Enterprise Development*, vol. 4, no. 4, IT Publications, London.

Hulscher, Wim & Fraenkel, Peter (eds) (1994), *The Power Guide*, IT Publications, London.

Hurst, Christopher and Barnett, Andrew (1990), *The Energy Dimension*, IT Publications, London.

IEA-PVPS, International Energy Agency Implementing Agreement on Photovoltaic Power Systems (1995a), *Examples of Stand-alone Photovoltaic Systems*, James and James Science Publishers, London.

IEA-PVPS, International Energy Agency Implementing Agreement on Photovoltaic Power Systems (1995b), *Use of PV Systems in Stand-alone and Island Applications*, Summary of Questionnaires, Internal Document.

IFC/GEF, *Photovoltaic Market Transformation Initiative*, Background paper (October 1996), International Finance Corporation, Environment Division, Technical and Environmental Department, and the World Bank, Global Environment Division, Environment Department, Washington DC.

IT Power (1997), *Photovoltaic Market Transformation Initiative*, ITP/97572 Project Announcement, May 1997.

IT Power (1996a) for the Asian Development Bank, *Solar Photovoltaic Power Generation Using PV Technology, Vol. I, The Technology*, Manila.

IT Power (1996b) for the Asian Development Bank, *Solar Photovoltaic Power Generation Using PV Technology, Vol. II, The Economics of PV Systems*, Manila.

IT Power (1996c) for the Asian Development Bank, *Solar Photovoltaic Power Generation Using PV Technology, Vol. III, Institutional Aspects*, Manila.

IT Power & Eurosolar for the European Commission (1996), *Final Report for Task 3, Financing Systems, PV for the World's Villages*, DG XII, EC Contract number RENA-CT94-0421.

IT Power (1994) for the European Commission, *Power for the World - A Common Concept, Study 1: Excellent Examples of Renewable Energy Applications in Developing Countries and Examples Not to be Repeated*, DG XII, contract JOU2-CT93-0027.

Jourde, Patrick (1991), *Integration of Solar Electrification: Pacific Island Examples*, 10th European Photovoltaic Solar Energy Conference, Lisbon.

Jourde, Patrick (1995), *PV Electrification and users in remote areas and developing countries. Lessons Learned: the Technology Ethnology*, 13th European Photovoltaic Solar Energy Conference, Nice.

143

Kabore, François and Durand, Jean Michel (October 1992), *PV Energy for a Sustained Economic and Social Development in the Sahel Region: The Regional Solar Programme*, 11th European Photovoltaic Solar Energy Conference, Montreux.

Karekezi, Stephen & Mackenzie, Gordon (Eds.) (1993), *Energy Options for Africa*, UNEP Collaborating Centre on Energy and Environment, Zed Books (publ.), London.

Kaufman, Steven (May 1994), *Solar Electricity for Rural Development: Experience in the Dominican Republic*, Energy for Sustainable Development, vol. 1, no. 1.

Kozloff, Keith (1994), *Rethinking Development Assistance for Renewable Energy*, World Resources Institute, Washington DC.

Leggett, Jeremy (ed) 1996, *Climate Change and the Financial Sector, The Emerging Threat - The Solar Solution*, Gerling Akademie Verlag, Munich.

Levitsky, Jacob (ed.) (1989), *Microenterprises in Developing Countries*, IT Publications, London.

Louineau, J P (1992), Zaire, *Review*, Issue 19, Department of Trade and Industry, London.

Louineau, Jean Paul, Dicko, Modibo, Fraenkel, Peter, Barlow, Roy and Bokalders, Varis, (1994), *Rural Lighting: A guide for development workers*, IT Publications, London.

Lysen, Erik (1994), *Photovoltaics in the South*, 12th European Photovoltaic Solar Energy Conference, Amsterdam, Netherlands, April.

Mandishona, Gibson (1994), *Empowerment for Africa*, ISES NGO Solar Energy Conference, South Africa, October.

Makukatin, S, Cunow, E, Theissen, M. and Aulich, H A (December 1994), *The CILSS-Project: A Large-scale Application of Photovoltaics in Africa*, First World Conference on Photovoltaic Energy Conversion, Hawaii.

Murphy, D P L, Bramm, A & Walker, K C (1996), *Energy from Crops*, Semundo Ltd, Cambridge.

Otero, Maria (1989), *Breaking Through: The Expansion of Micro-Enterprise programs as a Challenge for Non-Profit Institutions*, ACCION International, Washington DC.

Otero, Maria & Rhyne, Elisabeth (eds) (1994), *The New World of Microenterprise Finance*, IT Publications, London.

Owsianowski, Rolf-peter (April 1994), *Rural Electrification with PV in Morocco: A Market-Orientated Dissemination Approach in the Province of Kenitra*, 12th European Photovoltaic Solar Energy Conference, Amsterdam, Netherlands.

Padmanabhan, K P (1988), *Rural Credit, Lessons for Rural Bankers and Policy Makers*, IT Publications, London.

Panggabbean, L M (July 1993), *Photovoltaic Technology R&D and Applications in Indonesia, Country paper: Indonesia*, Expert Group Meeting on The Application of Solar Energy for Electricity Generation for Domestic and Commercial Use, Denpasar.

Pereira, O S, Moszkowicz, M, Hill, R and Stemmer, C E (October 1995), *Brazilian Reference Centre for Solar and Wind Energy: A Model for dissemination*, 13th European Photovoltaic Solar Energy Conference, Nice.

Preiser, Klaus, Kuhmann, Jerome & Parodi, Orlando (October 1995), *Quality Issues for Solar Home Systems*, 13th European PV Solar Energy Conference, Nice.

Previ, A, Ileceto, A, Belli, G, Buonarota, A, Gustella, S and Patona, R (1995), *Long-term Operational Experience at Vulcano PV Plant*, 13th European Photovoltaic Solar Energy Conference, Nice.

144

Rajabapaiah, P. et al (1993); Biogas Electricity - the Pura village case study in Johansson, T. B. et al (eds.), *Renewable Energy - sources for fuels and electricity*, Earthscan Publications Ltd., London.

Schaeffer, Loretta (April 1993), *Lessons learned from PV rural electrification in developing countries*, 12th European PV Solar Energy Conference, Amsterdam.

Schlangen, J & Bergmeijer, P W (October 1993), *PV Solar Home Systems in Lebak, West Java, Indonesia*, 11th EC PV Solar Energy Conference, Montreux, Switzerland.

Schweizer, Petra and Shrestha, Jagan Nath (September 1995), *Private Installations, a stimulus for PV dissemination? The case of Nepal*, In Search of the Sun, ISES conference, Harare.

Selling Solar, Financing Household Solar Energy in the Developing World (1996), Pocantico Paper No. 2, Rockefeller Bros Fund, New York.

Sheppard, Lisa, Richards, Elizabeth (1993), *Solar Photovoltaics for Development Applications*, Sandia National Laboratories, Albuquerque.

Smith, Julie (1996) Ed., *Enersol News*, North Chelmsford, USA, Winter.

Smith, Julie (1993/94) Ed., *Enersol News*, North Chelmsford, USA, Winter.

Stearns, K (1991) *The Hidden Beast: Delinquency in Microenterprise Credit Programs*, ACCION International, Washington DC.

Stöhr, M, Bischof, R, Dufner, H and Gregory, J (1995), *Power for the World: A Progress Report*, 13th European Photovoltaic Solar Energy Conference, Nice.

Thillairajah, Sabapathy (1994), *Development of Rural Financial Markets in Sub-Saharan Africa*, World Bank Discussion Paper No. 219, Washington DC.

UNDP (1996), *Energy Services for Small-Scale Energy Users in Southern Africa*, Project Ref. RAF/95/E01, New York.

UNDP (1995), *Zimbabwe - Photovoltaics for Household and Community use*, Project Document, Global Environment Facility.

United Nations Non-Government Liaison Service (1994), *The NGLS Handbook of UN Agencies, Programmes and Funds working for Economic and Social Development*, Geneva.

Van der Plas, Robert & Floor, Willem (June 1995), *Market-Driven Approach Can Illuminate Lighting Options for Rural Areas*, Industry and Energy Department, The World Bank, Washington DC.

Vegyan, P and Maureau, S (1995), *Energy Prepayment Schemes for a Sustainable PV Electrification on New Caledonia*, 13th European Photovoltaic Solar Energy Conference, Nice.

Wade, H; I H Sejahtera; T Ball (1993), Mission Report: *Evaluation of the Indonesian Photovoltaic Household Electrification Project*.

Wereko-Brobby, C Y & Hagen, E B (1996), *Biomass Conversion and Technology*, John Wiley and Sons, Chichester & New York.

WHO (1993), Working Papers for *'Solar Energy and Health'*, (SC. 93/conf. 003/8), World Solar Summit, Paris.

Wijesooriya, Priyantha (December 1994), Forum on Renewable Energy, *a Key to Sustainable Growth in the Asia-Pacific Region*, Hawaii, USA.

Williams, Neville (October 1994), *Financing Small PV Applications*, International Conference on Solar Electricity, Photovoltaics and Wind, Cairo, Egypt.

Williams, Neville (Nov./Dec. 1991), *'Solar Serendipity: Photovoltaic Rural Electrification in Sri Lanka'*, Solar Today.

World Bank (1993), *The World Bank and the Environment*, Washington DC.

World Energy Council (July 1993), *Renewable Energy Resources: Opportunities and Constraints 1990 - 2020*, London.

Wright, David L (December 1996), 'The pleasures and perils of donor consortia', *Small Enterprise Development*, Vol. 7, No. 4.

WWB (1993) *Building Strong Credit and Savings Operations*, Women's World Banking Manual Volumes 1 & 2, New York.

WWB (December 1994) *What Works: Achieving Policy Impact*, Women's World Banking Newsletter, Vol. 4, No. 5.

Yaquib, Shanin (December 1995), 'Empowerment to Default? Evidence from BRAC's micro-credit programmes', *Small Enterprise Development*, Vol. 6 No. 4, London.

Yaron, Jacob (1992), *Successful Rural Finance Institutions*, World Bank Discussion Paper No. 150, Washington DC.

Zaffran, Michel (ed) (July 1993), *Solar Energy and Health*, report prepared for the World Solar Summit, Paris.

Zaffran, Michel (April 1991), *Solar Refrigeration for the Storage of Vaccines in the Expanded Programme on Immunization in Developing Countries*, 10th European Photovoltaic Solar Energy Conference, Lisbon, Portugal.